Intersection: R

the future of mobility a...

traditional boundaries

Intersection: Reimagining the future of mobility across traditional boundaries

Alisyn Malek

INTERNATIONAL®

400 Commonwealth Drive
Warrendale, PA 15096-0001 USA
E-mail: CustomerService@sae.org
Phone: 877-606-7323 (inside USA and
    Canada)
    724-776-4970 (outside USA)
FAX: 724-776-0790

Library of Congress Catalog Number 2021948806
http://dx.doi.org/10.4271/9781468603958

ISBN-Print 978-1-4686-0394-1
ISBN-PDF 978-1-4686-0395-8
ISBN-ePub 978-1-4686-0396-5

To purchase bulk quantities, please contact: SAE Customer Service

E-mail: CustomerService@sae.org
Phone: 877-606-7323 (inside USA and Canada)
    724-776-4970 (outside USA)
Fax: 724-776-0790

Visit the SAE International Bookstore at books.sae.org

Chief Growth Officer
Frank Menchaca

Publisher
Sherry Dickinson Nigam

Director of Content Management
Kelli Zilko

Production and Manufacturing Associate
Erin Mendicino

Contents

Foreword

What does mobility mean anyway? I have been surprised numerous times that I needed to explain what I mean by it. I was just assuming that everyone would understand that I am talking about something related to transportation. But instead, some asked "Mobility, like solutions for persons with accessibility needs to get around?" Or as my Silicon Valley neighbors might say "Mobility, sure, just is another word for mobile computing."

Those examples tell us that mobility means different things to different people. But as different as those things might seem, they all apply to what my colleagues and I typically understand when we talk about mobility: the movement of persons or goods, and the creation of the ecosystem that enables that. And of course, providing exactly this to community members with accessibility needs plays an important role, and vehicles are also more and more becoming connected computers on wheels.

But even if my colleagues and I position mobility in the transportation field, still not everyone talks about the same thing. Some just use it as a different term for exactly that: transportation. Others would start to talk about a completely new ecosystem, consumer preferences, and the emotional aspect of owning—or actually not owning—a car. Those aspects and transportation might have as much in common as a 2022 high-end EV and a horse-less carriage from the early 1900s.

I would say transportation has a more utilitarian perspective, basically just getting someone or something from A to B. Mobility covers that as well, but also looks for instance at the emotional aspect of a car or a joyride where A and B don't matter much, but everything in between. In my view, transportation studies get us great insights into how much we travel and transport, what solutions are needed to make this as safe and as efficient as possible, or what the externalities of our transportation footprint might be. But I would maintain that only an understanding of mobility tells us why we cherish our next road trip or why we get a new car every 3-5 years just to have it parked in our driveway most of the time.

Now, we know pretty well where we are at with mobility, what makes it work, and what the problems are. But where are things going? What will mobility look like in the year 2030 or even 2050? To tackle those questions, Alisyn Malek did all of us a great service by asking an amazing roster of mobility thought-leaders to paint a picture of what the future of mobility could look like.

Consequently, Future of Mobility Explored by Industry Leaders, brings many and different perspectives together. There are renowned experts from the public and private sectors, NGOs and universities. They all share their individual perspectives where they see mobility is moving and what it takes to get things right. And in doing so, some discuss technology or infrastructure topics, others economic or societal aspects, and yet again others tackle regulation or public planning. An interesting observation is that, one way or another, they get to the point where mobility has a strong human component. For instance, one author points out "transportation is personal," or another expert states that the future of mobility is "definitely an experience … probably a vehicle." This confirms that mobility cannot be captured by a set of equations and models or something like that, especially as one author points out that he "unexpectedly tapped into something much deeper."

And still, the authors also debate who will create this future mobility that they all see shaping up as autonomous, connected, electric, and shared vehicles. Some make a strong point that it will take partnerships between the established players and newcomers in the industry. And equally much, others emphasize the importance of urban planning to make sure we will have livable cities for all people and not just efficient path ways for commuters. What's more, the authors also provide perspectives from many different regions of the world, covering the United States, Europe, and Asia. The contributions show that different regions require different solutions, but also that an international dialogue is necessary to capture the full potential of what future mobility holds for societies, economies, and the environment.

I do wholeheartedly agree with Alisyn that this is an exciting time, and we need to bring everyone to the table for the design, manufacturing, and operation of future mobility. It does require many different organizations and disciplines to interact in ways they may not have had to in almost a century. By reading this book, it becomes evident that we are not just talking about automobile 2.0 but a whole new mobility paradigm in the interest of safety, sustainability, and equity. We have a chance of our lifetimes, let's get it right.

Sven Beiker, PhD
Managing Director at Silicon Valley Mobility and
Lecturer at Stanford University
November 2021

1

Jonathon Baugh

Director of Experience - Slalom Consulting

For a kid from the Motor City, I've never been much into cars. Now that's not to say that I didn't enjoy my fair share of auto shows, Dream Cruises, Grand Prixs, and any number of other iconic "Detroit" automotive activities. I grew up with cars and car culture permeating my life so deeply that I'm convinced it's in my DNA. I've owned vehicles since I turned 16 because that is what you do in this part of the country—it might be a thrice-handed-down beater, but it's needed to traverse the Metro Detroit area. At 16 in the suburbs, my first vehicle was a ticket of admittance into early adulthood, letting me go to restaurants without my parents, shop on my own, and land my first job. For me, owning a vehicle has always been about utility, not about identity; my brand loyalty has switched a few times in my life based on what I needed out of a vehicle. While I am a "Detroiter" through and through—I have no problem driving 80 miles to get to work in the morning—few would argue that I'm any sort of a gear head or "car guy."

There were always others who were way more passionate about their vehicles than I, but even more so, I have always known that I did not want to work in the automotive industry. As an engineering-minded college student in Michigan, that made me a bit odd, and I realize that it is even bold to admit here, but it's the truth. Perhaps it is also what I need to get off my chest before I go further. Now, to be clear, I've spent most of my career in software and technology consulting and have worked for OEMs (Original Equipment Manufacturers, or automakers) and suppliers on everything from HUDs (heads up displays) to connected vehicles, DMSs (dealer management systems) to fleet management, and even vehicle remarketing. But every time I was invited into the inner circle of the auto industry as a consultant or even as a speaker at a conference, it was always as an outsider: a software designer, a "high-tech anthropologist™," or even as a connected product UX (user experience) expert, but never as an industry insider.

Perhaps it is with this license I'm going to say something that I don't think is said often enough:

The Automotive Industry Is Not Tuned for the Future of Mobility

In fact, the mobility industry will replace the automotive industry and be something wholly different than what we know now. The mobility industry will have a different value proposition to its customers. It will serve them differently with new offerings and different channels. There will continue to be new companies entering the market and a growing number of organizations, partnerships, and governments that would not be considered a "part" of the automotive industry of today. Already, we are starting to see this with technology-led startups, industry alliances, and various forms of public and private partnerships.

The entire business of mobility will look and feel very different from the automotive industry of the twentieth century. This isn't that controversial when you consider the changes sweeping the industry as of late. For example, very talented people have been working for many years to retool their companies and align for the next era in the industry. New organizational structures have been put into place, new leaders (including a fair number of outsiders) have been brought in, and companies are trying to change their approach to market through partnerships, investments, and all together new business models.

I am excited by these changes. Perhaps if this had been happening a couple decades ago when I was entering the workforce, I would have dove into the deep end and become a car guy instead of skirting around on the edges of the industry.

That said, I am a consultant. I strive to understand my clients' industries and see the big picture of where they have been and where they are going. I take lessons from other industries and apply them in new ways to solve the most challenging problems. This is the good part of being an outsider, you bring a fresh perspective most useful when you are trying to help a hundred-year-old industry pivot into the future. What follows are some of my larger observations and interpretations of trends around mobility that I've made over the last few years. Additionally, I'll provide my take on what changes will be required to meet the very different demands of the future of mobility.

Courtesy of Jonathon Baugh

Trends

For the last few years, it seems like everyone has a take on the "mega-trends" for the automotive industry. I won't spend too much time on these, but I do feel they are an important backdrop for what the future holds. In particular, I'll focus on how they will impact the customers of the industry.

Autonomous—Autonomous vehicles (AVs) will be more efficient modes of transportation in a number of aspects. While I believe that personal AVs will be desirable, they will have limited impact on how the industry serves its customers. Fleet AVs on the other hand are starting to make a mark. As far as mobility is concerned, this could impact municipal transportation services as well as privately held micro-routes. The infrastructure needed to design, operate, and maintain these fleets will be far different than is currently considered with "fleet" businesses.

Connected—With cell phone proliferation, having access to your mobility solutions at your fingertips is critical. Not only can this become a key differentiator in a company's customer experience but it is also a way for personalized mobility, which will be key when options are plentiful. Connected devices and apps are also a way for individual customers to traverse the countless features, apps, and even services available to them. Furthermore, thinking back to the growing fleets, managing these will require new levels of sophistication around software, data, and even privacy. Where I don't think the industry has a solid understanding is that connected isn't a 1:1 of vehicle to owner, but possibly a many-to-many relationship across numerous platforms—this will be a challenge.

Service based—Mobility is so much more than owning a car or having a bus pass. Already, we are seeing OEMs operating services that just a few years ago would have gotten you laughed out of a boardroom. Services, however, can be so much more than we are seeing today, more comprehensive service plans, tires-as-a-service, gas-as-a-service, and entire companies popping up, solely to operate services for larger industry players. As customers pivot away from buying or renting a vehicle, they will turn to alternatives focused on serving their needs across multiple platforms. Scooter proliferation across metropolitan areas in the USA is just one example of the ways that services will meet the customers where they are and flip some of the long-held paradigms of the automotive industry on its head.

Electrified—I am far from qualified to comment on the technological changes going through the Battery Electric Vehicle (BEV) space right now, but I personally felt the infrastructure challenges when a friend came to visit my house and needed to charge his vehicle for their return trip. How people plan their movements and, to a certain extent, support, carry-out, and pay for their movement is being rethought in real time. Who owns the recharging stations? How is their added and (currently) inconsistent demand considered on the grid? The automotive industry is not the sole decision-maker on how this will play out.

Business Model

These trends are not happening in isolation. They are occurring simultaneously along with thousands of other changes to the overall auto industry. The four mentioned above are just the tectonic shifts happening over entire careers, dramatically enhanced with a few significant earthquakes that happen seemingly overnight. These trends are areas that the incumbent companies and startups alike are betting on and investing heavily in. The changes will come in some shape or form, and the landscape will be redefined in ways we can't entirely fathom at this time.

Perhaps the biggest shift happening in the automotive industry is that of the overall business model. Recognizing that there are many different business models throughout the industry, they mostly focus on moving metal. That is to say that the OEMs primary business, dating all the way back to Henry Ford and all his innovations, is taking raw materials, manufacturing vehicles, and distributing said vehicles to consumers world-wide. Entire subindustries arose to enhance this process to control everything from expenses to ensuring predictability. To this day, much of the car purchase incentives systems are tied back to ensuring the metal continues to move in the planned and predicted ways. This will change.

Tomorrow's mobility customers won't tolerate being subservient to a system optimized for mass-producing vehicles. They don't want to be the last cog in the giant engine of the auto industry whose sole job is to consume, customers will demand that the industry serve them. They want a system that is focused on moving them and their goods, regardless of how often their needs change. If not, they will simply take their business elsewhere to a better fitting solution for their mobility needs than the time-honored tradition of buying a vehicle.

This shift from moving metal to moving people and goods is at the heart of the transformation required for the automotive industry to become the mobility industry.

To make this pivot, the auto industry must let go of some of the core tenants that it has built up over the last century and evolve to be customer focused. The history of the auto industry is one focus on inward improvements and efficiency, not outwardly on the customer. This must change. After all, the future of mobility will require more than just "any color, as long as it's black."

Control Is Out, Customer Is In

One of the largest (and likely most uncomfortable) shifts required for the future will be that of giving up the highly regimented controls throughout the industry. Historically, costs and supply chains have been tightly managed, production schedules were set months in advance, and labor management has been an ongoing historic struggle from the start. These controls (among many others) have been refined over decades to eliminate errors in both production and predictability. Command and control was the proven approach for orchestrating a highly complex ecosystem and producing the same results over and over. Collectively, the systems allowed the automotive industry to produce many tens of millions of vehicles per year worldwide and define much of the twentieth century, but will also prevent the auto industry from meeting the changing needs of mobility customers of the future.

Instead of controlling costs and timelines, the key for meeting customer mobility needs will be about flexibility and time to market. Automotive innovation was historically built on the model refresh lifecycle and companies measured progress over years, but that won't cut it anymore. Already, we are seeing that approach to innovation fall short of customer demands. Time and time again I have seen rigid and outdated production schedules dictate poor choices in feature prioritization for new vehicle production. In today's day and age of modern software devops and OTA updates, there is no reason to force mobile features on a connected vehicle platform to be locked in months before the physical vehicle is built. This archaic approach to planning means that customers get less of what they want, and that simply won't work for much longer.

In the new world, time to market will be far more valuable to customers. By understanding the differences in release cycles of software compared to manufactured vehicles, automotive companies could bring so much more value to their customers, but this will require letting go of some of the command-and-control processes embedded in the industry.

Partnerships Go Further and Faster

Another form of control that permeates the industry is the approach to innovation. Historically, auto companies have been quite protective of their innovations. They've cherished patents and IP (intellectual property), hiding their latest concepts and designs behind special secure studios and labs, afraid even their own employees could ruin their advantage with an illicit picture or two. Much of these innovations have been built upon iterative and incremental improvements that, when scaled across millions of vehicles, have profound impact on the bottom line. The problem is this innovation has been inwardly focused, not outwardly on the customer.

Mobility will require customers' diverse needs to be met through new approaches and solutions. Customers aren't willing to wait around for the big auto brands to figure out things in iterative steps, they want big change and they want it now. Net-new capabilities that are not a part of the automotive industry are already changing the landscape of urban and suburban transportation. These offerings are things that the incumbent auto companies aren't tooled to build. They don't have the intellectual property, the workforce, or even the business model to support ride-sharing or scooters as an example. Customers are signing up in droves for these new services, and all signs point to this as only the beginning of mobility for the masses. As such, we are seeing partnerships and investments from some of the biggest names in automotive happening more and more to drive faster to new and innovative offerings.

Production Shouldn't Be the Goal, Value to Customers Should

In perhaps the most cerebral of the shifts, the auto industry must realize that to become the mobility industry, production and consumption cannot be the goal—the holistic movement of people and goods must be. Customers want to live their lives, and your work product must serve them how and when they need it. It won't be as simple as making the next model of your flagship vehicles or finding new channels to distribute them. Instead, the future will be about learning what your customers want and finding the best ways to serve them. Their needs will be diverse. Mobility won't be solved by a single purchase or a single company. It will require a constantly changing ecosystem that evolves with them over time. As new technologies and services become available, customers will want to change their approach. Companies, governments, and other organizations will work together to understand those needs and determine the best ways to partner to deliver.

Producing vehicles that speak to the inner soul of customers will always be a desire of the automotive manufacturers, but the future of mobility will be so much more than vehicles (as many customers may never own a vehicle). The future of mobility will go beyond the cluster, beyond the driving experience, beyond the cabin, and even beyond the car all together. Your customer might not ever enter a vehicle. This is beyond everything that the last hundred years has taught this industry. The future of mobility will, in fact, be so much more than manufacturing, it will include trains, scooters, ride-shares, smart parking, charging stations, smart shipping, infrastructure, and any number of other services and platforms throughout an ecosystem that interacts and engages customers throughout their lives.

The future of mobility will be how customers and goods move freely and easily across a multitude of modes, platforms, and services, having far less to do with how the industry optimizes the manufacturing and distribution of vehicles and instead on how businesses, governments, and other organizations come together to create a holistic set of solutions.

About Jonathon Baugh

Jonathon has been designing and building meaningful digital experiences for over 20 years. Academically trained at the University of Michigan's School of Information and honed through multiple startups and several tours in consulting, his unique ability to fuse together people, technology, and business has proven instrumental in crafting modern customer experiences across companies of all shapes, sizes, and industries. This balance has helped his companies and clients create delightful interactions, meaningful engagement, and achieve significant company growth. Design Thinking supported by a strong backing of UX principles permeate Jonathon's approach to solving challenges in a pragmatic and human way. This methodology has proven invaluable in the diverse roles he has taken on, adding value on "everything but writing code." Jonathon now shares his passion and expertise with others through consulting, mentorship, and organizational growth. When he is not in the office, you can find Jonathon in the dojo or on his 10-acre homestead of wilderness heaven with his family (furry, feathered, and otherwise).

Courtesy of Jonathon Baugh

2

Courtesy of Geoffrey Boquot

Geoffrey Bouquot

Group Chief Technology Officer and
Vice-President of Strategy
Valeo

A few weeks ago, a friend of mine at Valeo showed me with great emphasis his family tree and asked me to look at it with close attention. I was not aware of his passion for genealogy and I tried to look for any pattern in the very dense labyrinth of first names and family names from the seventeenth century to the present. Looking at my puzzled face, he advised me to look at the locations of his ancestors, and there it was! His family tree offered irrefutable proof of the revolution in mobility driven by the introduction first of the railroad and then private automobiles in France. It was as simple as that. Looking at the tree's nodes and branches, you could see the transportation revolution that enabled people to live and raise their families far from where they had been born.

Mobility has a way of stirring all sorts of melting pots. As mobility became easier and more affordable, people's lives were no longer limited to their villages, and they could marry someone from the other side of the world. Advances in mobility are sources of human and cultural enrichment. And the reverse is true as well: social progress leads to improvements in mobility.

In contemplating my friend's family saga, the issue that we are passionately addressing at Valeo came to my mind more strikingly than ever: "Are we also, right now, at a watershed moment in the history of mobility?"

The technological revolutions that enable us to go somewhere else, to go further and faster, have already taken place. So, is that the end of the story? I seriously doubt it. When you're involved in mobility, when you work in it, what you're really doing is dealing with human beings and their lifestyles. The social upheavals we are witnessing today are inextricably linked to the transformations underway in our modes of transportation and how they are used.

So, what are these transformations? And what will they be like in the years ahead? They are all the transformations that enable people and goods not only to get from one place to another, but to do so BETTER. The major issues that manufacturers, and particularly Valeo, have to address are the ones involved in making mobility safer, greener, and more easily shared.

We are not alone in tackling them. Our operating environments are being defined by increasingly stringent policies deployed at the supranational, national, and local levels. These policies reflect the aspirations of people for a more sustainable world and CO_2 carbon neutrality commitments.

The acceleration of the introduction of electric cars is definitely an important sign of our times. Starting late last spring and in the space of just a few weeks, carmakers around the world unveiled their many plans for electrifying their ranges in the years ahead. Valeo has been anticipating this demand for more environmentally friendly mobility for a long time, to the point that we are now a vehicle electrification systems integrator. This means that we supply all the technologies needed to support the shift in powertrains, including e-motors, inverters (the brains of the system), onboard chargers, and charging stations.

This powertrain electrification revolution could almost be considered as over and done with in that it has already been embraced by manufacturers, public authorities, and other stakeholders. This view, however, ignores that we still have to meet at least three remaining challenges: (i) ensuring that the greatest number of people can access electric mobility, (ii) ensuring that the charging infrastructure networks exist and are at the right level of service, and (iii) finding alternative modes of mobility to private cars in cities where this personal mobility seems to be banned, even if they are electric.

As someone who lives and works in Paris, I can see these upheavals before my very own eyes on a daily basis. Cars are increasingly banned on certain city streets (such as the famed Rue de Rivoli), public transportation is overcrowded, the citywide speed limit will be lowered to 30 km/h from August 30, 2021, and on-street parking is more expensive and the number of spaces will be halved in the coming months. What I'm going through in Paris, others are experiencing or will experience in similar ways in London, Rome, Stockholm, Seattle, Copenhagen, and Tokyo. For more than two years now, I've been getting around Paris every day on an electric kick scooter. At first, my colleagues at Valeo thought it was just a fad or a whim. But I'm still using it. It says a lot about how modes of transport have adjusted to changes in the cityscape and urban lifestyles.

These days, mobility is no longer just a question of "how do I get from point A to point B?" It's about being able to provide solutions to an ever-growing number of increasingly complex problems. It's about dealing with vehicle size and bulk, parking problems, traffic flows, goods deliveries to people who don't travel around as much, system interoperability, congested public transport, and in-vehicle health.

These are all issues that reveal the extent to which mobility is changing. In response, Valeo has proactively developed a comprehensive portfolio of technologies to address them. These solutions are no longer just about cars, but cover all types of mobility, including those that have been described as "new" and "soft."

We're continuing to write history by asking ourselves the question: "How do we reinvent mobility in a digital world?" If we physically move around less, can't we come up with other ways of connecting people? One possible answer might be the virtual personal and business meetings held online during the pandemic lockdowns around the world.

Valeo is taking this to the next level, for example, by designing innovations that teleport, so to speak, an "immobile" traveler inside a real vehicle. Using an augmented reality display, he or she can see what's happening inside the moving vehicle, as well as all the real surroundings passing by outside. He or she can also interact with the driver.

In the past, mobility only went as far as the edge of the village. Today, it is limited only by state-of-the-art technology and our imagination. Who knows what we'll learn about mobility from future family trees? Perhaps what drives us at Valeo is the commitment to pushing back boundaries, so that we can bring people ever closer together.

About Geoffrey Bouquot

Geoffrey Bouquot joined Valeo in December 2015 as Group Vice-President of Corporate Strategy and External Relations. He is a member of the company's Operational Committee.

Since 2014, he had been technical advisor for industrial affairs in the Cabinet of Mr. Jean-Yves Le Drian, Minister of Defense.

Formerly, he served as a project manager in the Aerospace and Defense Unit of the French Government Shareholding Agency at Bercy. Prior to that, he was an adviser on purchasing and subcontracting at Latécoère and advisor to the CEO of the OCP Group.

Geoffrey Bouquot is a graduate of the Ecole Polytechnique (2005) and the Ecole des Mines, Paris.

Reilly P. Brennan

Founding General Partner
Trucks Venture Capital

It's Definitely an Experience, It's Probably a Vehicle

At various points in the last decade, we were promised a complete disruption of mobility due to automation. Or sharing. Or internet retailing. Or electrification.

Not one of these changed everything, but together they have done something more powerful: they pivoted customer expectations away from products and into experiences.

Today it's not merely enough to have a striking design on a showroom floor. A beautiful electric vehicle (EV) with a broken charging network is a failure. A smart car that doesn't integrate your phone seamlessly is dead on arrival. A hands-free driving system that disengages because it can't understand the human is a nuisance.

Experiences today are defined by the "vehicle and" something else, whether the vehicle is an automobile, bicycle, or personal aircraft. What a difference this makes for product designers: goodbye to Harley Earl, the legendary GM designer who often designed based solely on his whims and passions. Say hello to empathetic designers who consider their vehicles placed in a customer's life. The vehicle isn't good on its own, it is good because of the way the consumer uses it to experiences something meaningful.

The American architect Bernard Maybeck once said "the artist suspects that it is not the object nor the likeness of the object that he is working for, but a particle of life behind the visible." Transportation design has never been defined so perfectly.

This evolution will require new designers from new backgrounds to create the experience of a Porsche or a Trek bicycle or a Joby aircraft.

Years ago, at our little house in San Francisco, car designer Masato Inoue was over for dinner. Masato-san started designing traditional cars like the original Maxima sedan, but later in his career, he spent a lot of time and effort thinking about mobility systems, EVs, and a new way forward. His last big project before retirement was as chief designer of the Nissan Leaf EV.

In talking about EVs and systems, he said something quite beautiful: new systems bring new challenges, but they also bring new joys. One specific example he pointed out that hit home with me was the inherent stillness of an EV—with no vibration and very little sound, it can act as a rather effective tool to experience things near the vehicle. His cottage in Japan has a quiet forest road where he can drive his Leaf under 5 mph, turn the windows down, and hear everything around him. Thinking about EVs as a replacement for today's vehicle is therefore permanently flawed; it's just something else entirely, and there will likely be things about it we don't like, but many new things we do.

Transportation used to be defined by products in and of themselves. Today they are critical partners in new experiences. I don't think we're going back anytime soon.

About Reilly P. Brennan

Reilly P. Brennan is a founding general partner at Trucks, a seed-stage venture capital fund for entrepreneurs building the future of transportation. He has over 40 venture investments in transportation, including Roadster, Nauto, May Mobility, nuTonomy, Zendrive, AEye, Bear Flag Robotics, and Joby Aviation.

Reilly holds a teaching appointment at Stanford University, where he teaches each year in the School of Engineering and the d.school. His classes focus on transportation, design, and entrepreneurship. His influential newsletter Future of Transportation (FoT)is a radar for what's happening in transportation.

Prior to Trucks, Reilly was Executive Director for Stanford's automotive research program, Revs. He was a member of the Le Mans' winning factory Corvette C5-R program. His personal land speed record is 168 mph, behind the wheel of a Chaparral 2E.

4

Tiffany Chu

CEO & Cofounder of Remix
SVP at Via

How Wonderful

> We have to live differently, or we will die in the same old ways.
> —Alice Walker

To hop on a bike after a long day's work
Feel the breeze in our hair
And ride home with the week's cargo of groceries

To live a short stroll
From friends and loved ones
And drop by unannounced for an afternoon coffee
Just because we can

To walk—not rush—to the bus stop
Without looking at schedules or phones
Since we never wait more than
Five or ten minutes anyway

To hop on the subway for a last-minute errand
Strollers and babies and tote bags in tow
Knowing an elevator will always be there

To afford a cozy place to call home
With good opportunities nearby

Within a half-hour, or even a quarter
(In case this job doesn't work out)

To wheelchair down the street
Trusting the sidewalk won't end in crumbles
Arriving with dignity

For our kids to walk home from school by themselves
Or to walk our dogs peacefully in the evenings
Knowing we won't be stopped or frisked or be in danger
Because we belong here,
This is our neighborhood

To take the metro to the beach and stay late for the bonfire
To miss the ferry, catch the next, and still make the birthday candles
To throw a bike on the bus rack when rain starts to pour

To look around and discover
How we get to and from the places in our life, joyfully
Without diminishing our health or the planet's

Plus all the new ways to move
That we'll dream into reality
Because a diversity of good options for everyone
Equals an abundance of freedom for all.

The beauty of being connected to a place
To live our best lives

Whenever anyone asks me
About the "future of mobility"
These are the things I prefer to say.

About Tiffany Chu

Tiffany Chu is a designer and a planner. She is the former Chief Executive Officer and Co-founder of Remix (remix.com) and now Senior Vice-President at Via. Remix is the collaborative software platform for transportation planning used by 400+ cities globally and has been recognized as a Technology Pioneer by the World Economic Forum and BloombergNEF (New Energy Finance) for furthering sustainability and equity in the field. Remix was acquired in 2021 by Via, the leading transportation technology company powering public mobility systems worldwide.

Tiffany served as Commissioner of the San Francisco Department of the Environment and sits on the city's Congestion Pricing Policy Advisory Committee. Previously, Tiffany was a Fellow at Code for America, the first UX hire at Zipcar, and is an alum of Y Combinator. She has been named in *Forbes'* 30 Under 30, LinkedIn's Next Wave of Leaders Under 35, Curbed's Young Guns, and featured at SXSW, Helsinki Design Week, *The New York Times* Cities for Tomorrow Conference, and more. Tiffany has a background in architecture and urban planning from MIT.

5

Jordan Davis

Executive Director
Smart Columbus

M y favorite vision of the future of mobility is, of course, one that offers ultimate convenience and experience. Let's imagine… It's a Tuesday morning and my husband and I hail a ride that arrives on time at our house to pick up our family. The vehicle is autonomous and gives us additional family time together on the way to dropping them off at school. We are all riding facing each other and talking through the day ahead, not worrying about other drivers on the road, missed turns, or unexpected road closures. When we drop off the kids, my husband gets on a scooter to head into work, I stay with the vehicle, pop up a desk in the interior cabin, and do a quick telemedicine appointment with my doctor en route to the Hyperloop station. When I arrive at my destination, the vehicle pulls in over a wireless quick-charging pad to let me out right in front of the entrance to the station. I end my ride in the app and use the same interface to pay and board the Hyperloop seamlessly. I take a 30 min tube shot from Columbus to Chicago for a few business meetings and am back in time for school pick-up at the end of the day.

And, oh, by the way, all of the small, time-consuming errands are done for me while I've been gone. My dry cleaning has been picked up, groceries delivered, and dinner at the house.

This hyper-individualized, interconnected transportation experience sounds absolutely amazing to me! This vision of my normal future-Tuesday incorporates some of the best mobility visions out there. And yet as desirable as it may be, the systems change required to bring this particular vision to reality is more complex and dynamic than the most infamous stories of technology disruptions in history.

A little over a century ago, the car was introduced mainstream, and in the following decades, paved roads, street lights, and ultimately highways were built out across the country. These physical infrastructure changes facilitated the growth of the automotive industry and, in turn, impacted every aspect of life from the migration of society, physical design of our cities, individual safety, and economic potential of industry. The same will occur as a result of this next transportation revolution. Might we get rid of street signs altogether? Will midsize cities become suburbs of megacities? Will we no longer need as much parking? The physical infrastructure shifts will undoubtedly deeply alter our communities and society. But most notable might be the simultaneous and faster change that is occurring in the virtual world as a result of individualized user experiences, new interoperable systems, and data sharing.

I've spent the past five years as part of the team implementing projects and programs as a part of the winning US Department of Transportation's Smart Cities Challenge. Through this work, I led the deployment of Ohio's first self-driving vehicles on public roads, spurred the fastest EV market in the Midwest, supported the advent of scooters, helped our region propose a Midwest Hyperloop route, and supported the city of Columbus in their attempt at developing a common payment system and integrated data-sharing platform. I touched all the seeds of what could make this future-Tuesday possible.

Spoiler alert, this won't be our reality in the next decade, but I believe our micro-decisions today will have macroimplications tomorrow. Said differently, the steps we take this decade will shape what our future looks like in the next one. If we want a future-Tuesday somewhat comparable to the one I laid out, we need to start partnering and working together for real, both inter-sector and cross-sector.

For example, the instance where the autonomous vehicle drops me off in front of the Hyperloop station while simultaneously idling on top of an inductive quick charge pad, to top off the battery will not just be a feat of technology but also one of multi-stakeholder partnerships:

1. Industry will need to agree on charging standards that all models use.
2. Transportation providers and payment systems (Uber, Lyft, Hyperloop, Mastercard, VISA, and banks, etc.) will need to create an easy and integrated method to schedule and pay for all different pieces of my trip seamlessly.
3. Permissive policy will have to be set by the real-estate owner prioritizing the user experience in selecting that convenient location.
4. Investment from the utility, and, in turn, likely the ratepayer (i.e., you), would be required to ensure adequate power at that site.
5. A dynamic financial incentive structure would be needed to ensure not only that the chargers are used and maintained properly but also that they remain equitable to all rider trips, not just those who can afford it.

This is just one very small touchpoint in the future transportation network that is being defined today as industry and cities are on the brink of setting precedent at scale for how the transition to new mobility solutions will occur.

My hope is that we in America focus less on what company will get there "first" but rather on those who can form the most partners in the process, and how all Americans will benefit from this transformation.

About Jordan Davis

Jordan Davis serves as Executive Director of Smart Columbus, a public-private inno-vation lab that advances new and next solutions at the intersection of technology and community good. Jordan was a part of the founding team of Smart Columbus when the city won the US Department of Transportation's Smart Cities Challenge in 2016. Through her work with Smart Columbus, she has overseen one of the country's first self-driving vehicle deployments, formed partnerships with over 100 different orga-nizations representing $720 million of aligned investment, and directed Columbus' effort to increase electric vehicle adoption by nearly 500% breaking world records for EV education and leading the Midwest in market growth.

Jordan has been fortunate to speak to audiences around the world about her pioneering work and has been honored as one of the "Top 25 Government Doers, Dreamers, and Drivers" by Government Magazine in 2020, "Top 100 Influential Young Executives" by *American City Business Journals*, a "Rising Star" by *Automotive News*, and one of ten Mercedes-Benz Future of Mobility Fellows in the world.

Courtney Ehrlichman

Head of Strategy
Panasonic Smart Mobility

A beautiful orchestration of code, lasers, and data, driverless vehicles offer us the promise of harmonizing our congested streets while also making them safer, rescuing us from our own human error and fallibility.

Software is being written so that, at the tap of a button, these self-driving cars will know where to pick up and deliver your groceries, when it's time to get you to work, or fetch you from the club on a wild Saturday night.

Parking will be a thing of the past, not to mention paying for it or for parking tickets. Say goodbye to all that time wasted circling to find the perfect spot.

As these sensing, artificially intelligent vehicles gather data and create beautiful 3D maps of our world, they will also begin to be much more than a vehicle that takes us from a to b.

Ladies and gentlemen, we are at the frontier of a new Mobility Revolution. **Code is the new concrete. Lasers are the new eyes on the road. Electricity is the new fuel**.

We've been promised that electric, driverless vehicles have the *potential* to save lives, cut down on emissions, and solve a panacea of our mobility woes.

They better.

In 2013, I coordinated a 33-mile driverless car ride for the Chair of the House Transportation and Infrastructure Committee and the then-secretary of PennDOT in Pittsburgh. Since that time, two of my dearest friends have been killed by cars.

In December of that very year, Andy Fisher, a doctor exercising the Hippocratic Oath after he was involved in a 50-car pileup on the snowy Pennsylvania Turnpike.

Dr. Fisher stepped out of his car, leaving his wife and two children safe and warm inside it, to tend to the victims of the crash, only to be struck by a vehicle going far too fast. He died on the scene.

Two short years later, Susan Hicks, a professor at the University of Pittsburgh, was killed by urban traffic violence. She was simply stopped at a light in the heart of Pitt's campus, on her way home from work. A drugged driver rear-ended the vehicle behind her, which thrusted forward and pinned Susan between two cars. She was killed instantly.

The impact on the community had a ripple effect and these lives were not lost in vain.

Approximately 40,000 people die in traffic violence every year in the US—1.3 million people every year around the world.

At the time of both of these tragedies, I was working as Deputy Director at the USDOT's National University Research Center for Transportation Safety at Carnegie Mellon University. There I worked with the Robotics Institute to address real-world transportation problems. And while the loss of Andy spawned the use of A.I. to detect snow conditions on the road, I spun the technology out of academia, co-founding RoadBotics. Today, RoadBotics focuses on infrastructure assessment in more than 250 cities around the world. In fact, we developed countless technologies in response to real-world problems, including driverless cars.

Despite the hype, despite the trough of disillusionment, despite the failed deployment timeline, driverless cars give me hope. We need them to fulfill their promise. We need them to save us from ourselves. But how can we make them work for all of us when we don't even know what that means yet?

Dystopian Concerns

First, let's talk about those parking tickets. NYC raked in $545 million in parking ticket revenue in 2017. Parking tickets are annoying at best, but they fund our cities. The downward trend we are seeing in car ownership right now coupled with cars that can drive themself means the need for parking, especially on-street parking, dramatically drops.

So, if the cars aren't parking, what are they doing? They are roaming, moving around all around on the street all the time. Waiting for someone to call them. Why is this bad? Emissions, congestion, and safety.

Our national and state roads funded by the gas tax, face a triple, but very necessary, threat—more efficient vehicles, a trend towards fully electric vehicles, and our new work from home patterns.

Essentially, the way we fund infrastructure in this country is about to be toast.

But what about these sensing vehicles making beautiful maps of our roads and the world around them? They rely on an infrastructure system that is well maintained and upgraded to support flows of data and connectivity.

Are these driverless cars really going to solve all of our problems? The more we take a step back from the hype and recognize that these vehicles are an amalgamation of algorithms designed to replace human drivers, aren't they missing a major piece of the equation—humanity?

What if we, the people, demand that this new mobility revolution also be orchestrated with our values, tailored to our community's needs, like affordable housing and economic access?

How can we guarantee that data will be balanced with privacy, that implicit bias won't be trained right into the algorithms?

In my world, I hear over and over again from cities that they want data from new mobility companies. They just need to know where people are coming from and going so they can optimize transit stops and utilize this data for other city purposes. But who owns this data? Don't I own my own origin-destination data? Do I want either of these entities, the companies or the government to have access to it? Do you? Do we have a choice?

We have to understand, now, that if we just let technology drive this revolution, we, the people, will not be part of the equation.

So here is our opportunity to engage.

We have a lot of data that we are just giving away. We are letting companies develop their products on our roads, in the public right away, with us as the crash test dummies.

What if we flipped this concept on its head? What if these companies shared what they are learning from our data with us? What if we could create value from the data exhaust of these companies?

What if we are missing the point completely with origin-destination data? Wouldn't it be more powerful to know where people are NOT going because they can't get there? What if, we, the people, gather our own data and turn around and sell that to the industry? Forming community data trusts, while also providing powerful insights to our local governments for planning purposes. It is possible.

About that parking. What if cities found powerful ways to reuse the land that is freed up from parking to address community needs, like affordable housing infill development or fresh food markets?

How do we make sure we are not creating a river of cars and freight that are constantly roaming around our streets? What if we limit them to certain streets and create streets that are dynamic across the course of the day, allowing freight at intervals through the day and becoming pedestrian/bike only at other points of the day?

So are cities and communities even equipped to harness this potential future?

In terms of safety, the Federal Government is moving tentatively in its framing out of its regulatory process. States have created a patchwork of legislation that

So how do we, the people, trust the beautiful orchestration of code and machine learning that is on our streets, testing right now? Our first automotive safety standards were established in the 60s with the help crash test dummies in automotive crash tests. The same needs to happen for the software that is being incorporated into these cars. We need to see policy REQUIRING third-party software testing and validation. That way we can avoid incidents like the one we saw with the death of Elaine Herzberg in Arizona or in the many Tesla Autopilot crashes. Otherwise, these vehicles are not going to be any better than humans.

We need to ensure our participation in developing the mobility future that matches our values and creates a society that we want to live in. To do that, one could say, we need to be in the driver's seat.

About Courtney Ehrlichman

Courtney Ehrlichman is a strategy and innovation leader in transportation. As Head of Strategy at Panasonic Smart Mobility—the innovation center at Panasonic North America—she plays a key role in shaping Panasonic's path in the emerging world of new mobility. In 2016, she co-founded RoadBotics, a Carnegie Mellon University Robotics Institute spin-off that uses artificial intelligence to power infrastructure assessments in over 250 cities globally. Prior to that, Ms. Ehrlichman spent over 10 years solving real-world problems at Carnegie Mellon University as Deputy Executive Director of the Traffic21 Institute and two National USDOT University Research Centers. Today, she proudly serves on the boards of Partnership for Advancing Responsible Technology and Intelligent Transportation Society of PA and on PA's Autonomous Vehicle Task Force and PittCyber's Task Force on Public Algorithms. Ms. Ehrlichman is a *TEDx* speaker and has been a guest on NPR *Marketplace*, Vox's *Answered*, the *Autonocast*, *The Mobility Podcast*, the *PolicySMART* podcast, and *PBS Newshour*. As a single mother, Ms. Ehrlichman earned her graduate degree from Carnegie Mellon University.

Elaina Farnsworth

CEO
The NEXT Education

As CEO of The NEXT Education, I'm proud to lead an extraordinary team of educators, innovators, and pioneers in the field of Mobility Education. We are dedicated to developing the best certification and training solutions for the workers of today to transform them into the new mobility leaders of tomorrow.

How did I get here? I didn't get here overnight. Instead, I started on a journey many decades ago that I could not have planned. With serendipity on my side and a bit of courage in my pocket, I found myself embarking on a journey that has led me to a wonderful place—to a career that I love, fulfilling my passion to help people become a part of this new world of Mobility.

I grew up in a small rural town where technology was sparse and the internet was just entering the scene. I had a love for the huge boxes with typewriters attached that were called computers. Although many folks at that time did not know what to do with them, as a teen I frequently found my way to the "computer lab." The lab with one machine in the dark corner of the Home Economics room. It was here, in my small school, that I found my love for technology.

Fast forward many years later, I was the only female in my master's degree program at Webster University. Computer Science wasn't a very popular field yet, and certainly not for women. One of my favorite stars growing up was Gillian Anderson, who played Agent Scully in the 1990 TV Series "The X Files." I didn't think much about blazing the way as a female, but if Agent Scully could do it, so could I! I later learned that the "Scully Effect" became a studied phenomenon that inspired many women to be a part of science and technology.

Post grad school, my early career took me many places across the US. I became part of the exciting world of technology startup during the dot-com days of the 1990s. I was fortunate enough to work with numerous corporations, universities, government programs, and worldwide organizations. In each of these roles, my team and I developed new applications that supported emerging technologies in transportation and emergency response. We used software, hardware, communication, and security to innovate. These concepts, protocols, and technologies are still used as the foundational systems for what we know today as Connected Vehicles (CV) and Intelligent Transportation Systems (ITS). Each of the applications we built were innovative and new.

Each win we celebrated. We were excited for the changes we were creating in this world. We were becoming a part of the technology revolution! We failed as many times as we succeeded, if not more. One thing was constant ... We needed people. In every equation, in every application, we realized that the success of the technologies we produced were based on the people's ability to use them. How can people be equipped to succeed with new technology and innovation? Invest in their success. Train them. I have always believed in investing in people. If we invest in people's skills, we invest in the development of the workforce that will grow the industry.

I had been happy with my success as a technology entrepreneur. One of my fruitful ventures led me to Detroit Michigan. It was here that I would be propelled from technology entrepreneur to mobility maven. Detroit is a wonderful city. It is rich with history and full of world-class firms investing in automotive and in technology. As I began to meet leaders and wonderful advisors in the metro area and beyond, I felt compelled to capture their experiences to share with others. In the spring of 2010, I assembled a group of partners and created a company specifically to train people in new mobility technologies that were emerging in automotive. New Mobility is not just a word. It is a system of systems. Many many parts working together to make the movement of people and things better, safer, and more equitable. As we spoke to experts, we realized that we were witnessing a phenomenon like no other. A change that transportation has not seen since the transition from the horse-drawn carriages to the invention of the modern car and the assembly line. Cars are now becoming connected. Vehicles are beginning to talk to other vehicles. Vehicles are beginning to talk to traffic lights and infrastructure, and vehicles are beginning to talk to other things. Vehicles are becoming a mobile device, and twenty-first century infrastructure systems are emerging. We needed to train people to use them, to build them, to maintain them. People need to have multiple skills to understand the complexities and the inner workings of the systems. This is the future!

For the next ten years that followed, my team and I developed award-winning certification programs for people. I spoke at hundreds of events, and we spent hundreds of hours researching and aligning curriculum to certification programs. We partnered with universities, and we searched the globe for specialists in connected vehicles, autonomous systems, intelligent transportation systems, and cybersecurity. We partnered with industry greats such as SAE International, CompTIA, the Institute of Transportation Engineers (ITE), the Connected Vehicle Trade Association (CVTA), the Center for Automotive Research (CAR), and other wonderful organizations to launch some of the first certifications in the Mobility Industry. We realized the importance of cross-functionality. We realized that there are so many skills needed.

This decade, the twenty-twenties, has been a remarkable time. There are few times in history where we have seen such change in our world and in the world of transportation. Where there is change, there is opportunity! Today I still believe in the power of training. I believe more than ever that we need to equip people, regardless of their background, with the training to succeed in the future.

The global pandemic forced the world to use technology in a different way much like the internet forced a change. The world connected online like never before, and today's advancements in transportation, with connected cars, autonomous vehicles, and unmanned aerial vehicles, are pushing the limits of innovations. Our world is changing at lightning speed. The advancements in technology and new mobility have caused a significant opportunity to transform our workers of the future to get jobs in this exciting field.

As we enter the future of transportation, we're becoming part of a world where convenience rules the day. We no longer need keys to start engines, we can voice command our way through making phone calls, selecting music to drive to, take-out orders for dinner, and directions from A to B.

On the other side of all of this convenience are the people who keep it running. Maintaining technological complex systems is where the crossover between engineers and technicians takes place. The problem is that combining those skill sets is uncharted territory, particularly when it comes to education. This new technology frontier calls for a revamped system of training to prepare the next generation of hybrid "automotive technicians" slash "software gurus" for the workforce.

I look to the future with hope and excitement as I see opportunity for a more equitable, more sustainable world. I see a world where those who need transportation can have access to it through these technologies. People who never had access to transportation can have self-driving vehicles accessible to get them where they need to go. As the need for more contactless delivery increases, I see how a more sustainable transportation structure will benefit communities. As autonomous vehicles on the ground or in the air enter the scene, how will our worker network need to adapt? What is the solution? Obviously, we must close this growing gap to train people in this new world that will include concepts of new mobility, smart cities, and intelligent infrastructure.

In order to move forward, we need more people with a knowledge base and skill set in the mobility industry—people with the passion and the purpose to change the future. But how? The precedent would suggest taking a trip to top colleges and recruiting a whole new team because, in the past, we've had generations between the changes taking place in technology and have been able to solve the learning gap by slowly bringing in fresh faces across a few years. But current advances in digitalization are moving too fast for that. It's impractical today to find an entirely new team. Not only can companies not justify a complete employee turnover, they also can't afford it. Fortunately, those very same people, technicians, installers, and software engineers in the field, have provided us with the solution: learning new skills. They've been retraining and upskilling themselves. They've sought out conferences, watched videos, experimented, and learned by trial and error until they found the skills they needed. Then they grew, worked their way up that corporate ladder, and began training others to follow in their footsteps.

And that's the solution. Finding suitable, formal training/reskilling programs for people and companies is the realistic response to close this gap. This new way of learning will require an overhaul in the way workers are trained. There's an ever-growing gap to support all of the growing workforce needs that will be required in the new era of connected vehicles, self-driving cars, and twenty-first-century infrastructure.

What is the good news? We will have new opportunities to grow these industries. How? Through growing people to be agents of change for the future. And that is where my passion lies. I encourage people to find the best education for them. I encourage companies to form partnerships with formal further education providers, like *The NEXT Education, SAE International, Universities,* and other great organizations in the industry. I encourage us all to work together to find the resources for retraining and upskilling current employees.

I have been blessed to find dedicated, accomplished, and seasoned instructors who have gained the skills and are wholly invested in sharing that knowledge can relieve some of the pressures felt by management teams, HR representatives, and CEOs that feel responsible for their employees. I work with professionals from an international range of industries and are prepared to take on this challenge of training. My team and I are hands on, ready to roll up our sleeves and get our hands dirty to create actionable solutions to close that ever-growing skills gap together. In this ever-evolving field, new job opportunities arise each day. We need more workers with all backgrounds and different skill sets. There are no limits. Mobility is here.

About Elaina Farnsworth

Elaina Farnsworth is the CEO of The NEXT Education. She is passionate about the need for knowledgeable leaders, educated employees, and skilled tradespeople in the fields of connected and autonomous vehicles, new mobility, cybersecurity, and twenty-first-century infrastructure technologies and deployment. Her work in these fields continues to equip others with the skills and knowledge to advance their careers in an ever-changing world. An acclaimed speaker published writer and thought leader in the connected vehicle and cybersecurity industries, Elaina has been recognized as *Industry Era*'s 10 Best Women Leaders of 2020, 2018 Top 10 Influencer for North American Automotive Suppliers, 2015 TechWeek100's Top Tech Leaders in Detroit, a Crain's 40 Under 40, and 2015 *Corp Magazine*'s Most Valuable Professional among others. She serves as an advisor to the US Army GVSETS Cyber Industry Chairperson and the Michigan Automotive and Defense Cyber Awareness Team (MADCAT). She was also appointed Director of Global Communications for the Board of Directors of the CVTA and sits as an advisor to other educational institutions and nonprofit organizations.

She lives in the Detroit metro area with her two children, Brooklyn and Kyle, her husband Don, and her dog Gigi.

8

Valerie Lefler

Founder and Executive Director
Feonix - Mobility Rising

My faith has always been a significant part of my life, but never more than in my current role as the executive director at Feonix - Mobility Rising. I believe that my life's calling is to improve mobility for those underserved in the current transportation landscape because a lack of mobility has significant impacts on the health, economic opportunity, and educational options for our communities.

When I think about the future of possibilities for mobility 20 to 50 years in the future, I am beyond excited about the potential for autonomous vehicles (AVs) and smart mobility to impact those most in need. Yet it is sobering to think that it's been 31 years since the Americans with Disabilities Act passed and we still lack basic sidewalk infrastructure in most of our communities surrounding our transit stops, ride-hailing services rarely have accessible vehicle options, and to book rides on most paratransit systems require calling 2-5 days in advance, and often waiting on hold for 20 minutes or more.

As we strive for a better tomorrow, I believe equity must be a priority and that there are many "low tech" advances that can easily be made today that will improve the quality of life for millions of Americans in the next 50 years. For example, booking a ride via app, or coordinating transit services online are still major wins that must have dedicated effort for implementation for the full benefit to be realized by society.

In our Michigan Ride Paratransit project, sponsored by the Southeast Michigan Transit Authority and Michigan Department of Transportation, we deployed a simple app for paratransit customers that enables them to request rides via an app versus calling in on the phone, and it was life changing for so many. One passenger reported that just the time he saved sitting on hold, gave him over 200 hours a year of his life back that he normally would spend sitting on hold. Another passenger was a nurse who was supporting her aunt as a caregiver, and without the app to book the rides, she would not have been able to provide the level of care and support her family needed during the pandemic. It is these minor adjustments to the mobility technology landscape that I believe have the biggest impact in creating an equitable future.

While the possibilities and excitement around the future is growing as we think about AVs taking us to work each day or delivering our pizza, we must affirm and acknowledge our commitment to equity, and using new technologies, not just solving issues for parking and traffic congestion but addressing decades of systemic barriers in access.

For example, research indicates that 5.8 million Americans miss healthcare appointments because they simply lack the transportation to get there, but how many of those appointments would have been successful if an autonomous vehicle equipped with telehealth equipment showed up at their door? The "high tech house call" connected with an entire suite of healthcare professionals accessible from their driveway and a seat on the vehicle. If we are going to sustainably fund and optimize value in our future transportation ecosystem, I believe we will need to engage healthcare, education, and major employers to establish new use cases and revenue models.

As an industry, we also need to look at the upstream equity barriers in the mobility business ecosystem. For example, I believe we need to establish tech start-up incubators in the transportation innovation space that promote BIPOC entrepreneurs and entrepreneurs with disabilities. And further upstream, we need to create more STEM education programs that focus on exposing underrepresented students to the transportation industry—engaging them early on in their career pathway.

These early-stage investments will support a future, such that in 20 to 50 years we have the best workforce and that we're thinking inclusively about our nation, as our leaders in the industry reflect the communities we serve.

While it's not as exciting to talk about after school programs, ticketing options for individuals who are unbanked, or accessible vehicles as it is new patents, sensors, or algorithms in most transportation executive conference rooms, they are the conversations that I believe will result in establishing a firm and equitable foundation for the future of mobility to impact and improve as many lives as possible.

About Valerie Lefler

Valerie Lefler is Founder and Executive Director of Feonix – Mobility Rising. Ms. Lefler is an international expert in rural transportation, accessibility, and mobility as a service. In just its first three years, Feonix has launched programs in seven states with notable partners, including the National Aging and Disability Transportation Center, Easterseals, the AARP Office of Driver Safety, Toyota North America, Centene Corporation, and the Michigan Department of Transportation. Lefler was featured by *Smithsonian Magazine* as one of the "Top 9 Innovators to Watch" and has also been highlighted in publications such as the *New York Times*, *NPR*, and the *Christian Science Monitor*. Valerie graduated with distinction from the University of Nebraska–Lincoln with a degree in Business Administration after studying International Economics abroad at the University of Oxford. She also received her master's degree with honors in Public Administration with an emphasis in Public Management from the University of Nebraska–Omaha. She and her husband Joe now reside with their two children in a small town outside of Lincoln, Nebraska.

9

Wolfgang Lehmacher and Mikael Lind

Wolfgang Lehmacher
Global supply chain and technology strategist

Dr Mikael Lind
(Adjunct) Professor in Maritime Informatics, Chalmers University of Technology
Senior Strategic Research Advisor, Research Institutes of Sweden (RISE)

Building a Fluid Goods Mobility System

A Call for Moving Beyond Data Sharing for Supply Chain Visibility

Even after decades of working in the goods mobility industry, we continue to be fascinated by the ability of networks to deliver all types of merchandise in enormous quantities to almost any point of destination across the globe. The secret mechanism enabling this outcome is the self-organizing nature of the multi-stakeholder goods mobility ecosystem composed of autonomous players operating along the chains across different cultural environments and various legal jurisdictions. Ensuring this high level of delivery performance in the future requires that the mobility industry reengineer its practices and digitize its processes as a collective effort with contributions from all stakeholders involved.

The goods mobility sector is characterized by players that act in strong co-opetition, competing in some aspects of the business while at the same time collaborating in other areas. The broad landscape and the dynamics of this situation are more efficiently managed with architectures and applications that provide visibility based on data sharing across the global, regional, and local networks. Establishing this level of visibility has become an increasing aim in the industry over recent years. However, visibility is not an end in itself but needs to result in network-wide synchronization and integration to unleash productivity reserves inherent in the resources and infrastructures deployed. Applying principles of the appointment economy in form of slot management will help to make the best use of a digitized goods mobility system and ensure the fluidity of the system.

The willingness, however, to invest in knowledge, capabilities, and processes supporting this vision has traditionally been limited. Much of the previous improvements have produced suboptimal results by taking only smaller steps rather than bigger leaps that challenge existing patterns of behavior and traditional business models. The verdict is out whether current pressures to innovate, exerted by customers and regulation to mitigate climate change and other risks, will bring broader change across mobility networks through increased investments in talent, digitization, synchronization, and integration. The high freight rates and positive financial results, particularly in the maritime industry, should help to finance some of these investments into the future.

We have been engaged in the goods mobility sector with the goal of making a difference by supporting the efforts of the goods mobility industry towards a connected synchronized and integrated goods mobility ecosystem. Wolfgang, with a strong emphasis on supply chain and logistics innovation, and Mikael with a focus on intermodal supply chain digitization. During the last ten years, Mikael has been deeply engaged in the maritime sector, becoming accredited as the first (adjunct) professor in Maritime Informatics as a collaborative endeavor between the Research Institutes of Sweden (RISE) and Chalmers University of Technology—a collaborative initiative contributing to awareness and knowledge building. Wolfgang has leveraged his substantial experience acquired through working on a global scale as Director for Supply Chain and Transport Industries at the World Economic Forum, as Global Logistics Lead at the strategy firm Corporate Value Associates (CVA), and as President and CEO at GeoPost Intercontinental to drive and pursue strategic investments and partnerships to upgrade capabilities. Mikael is the co-founder of Port Collaborative Decision Making (PortCDM), a concept inspired by Airport CDM (A-CDM) now gaining interest for other types of (multimodal) transport nodes as well. He has also been one of the initiators and leaders for the emerging discourse on Maritime Informatics (www.maritimeinformatics.org) empowering practitioners and academics at a global level to join forces on digitalization with the aim to achieve enhanced coordination and synchronization in maritime operations. Wolfgang is deeply involved in improving efficiency, performance, and the sustainability of goods mobility systems at the crossroads of innovation and operation, instilling change in large corporates while supporting asset owners and innovators to transform the industry towards digitized practices and models.

Some time ago, we joined forces to explore and drive an industry agenda that puts emphasis on how digitalization can empower stakeholders and improve operational performance through situational awareness and supply chain visibility as the foundation for better decision-making and automation. Our joint work is focusing on multimodal goods mobility actors operating across primarily maritime supply chains, organized end to end from shippers to receivers by beneficial cargo owners, carriers, and logistics service providers. Our shared aim is to augment capabilities and performance across networks through increased synchronization and integration to improve asset and infrastructure productivity. In the process, we have discovered that there is a lot to learn from studying practices applied within and across different nodes and modes on supply chain optimization across different regions and cultures.

Informed by the pandemic-induced disruptions and the resulting volatility and uncertainty across the global economy amplified by shifting consumption and trade patterns, we formed the opinion that many supply chain networks have reached their limits. This is particularly evident in the maritime industry, evidenced by the port congestions that have regularly occurred in different geographical regions. They cause ripple effects across local, regional, and global supply chains, resulting in delays at the point of destination and the various stages along the chains. The aggregated disruptions have not only caused major imbalances in equipment and a spike in container freight rates, but they also significantly reduced predictability and, consequently, the ability of the various actors to plan their supply and inventory and ensure that sufficient capacity is available.

The recent regular congestions that have emerged at Long Beach and Oakland are causing up to three weeks of ships waiting outside the ports with unnecessary utilization of the earth's resourcses.[1] The congestions also amplify the lack of truck drivers in many parts of the world. The supply chain system cannot afford to have trucks and trains waiting at different sites close to ports but also at strategic inland gateways, like Chicago, for ships to finally make their port visits. Beneficial cargo owners struggle to plan for such a situation. In the meantime, the disruptions ripple across the world. The current relief at Long Beach and Oakland is not the result of improvements of operations but another disruption caused by the pandemic-driven governmental measures that struck some ports in the south of China, which forced them to operate at significantly reduced capacity utilization and performance levels.

An unrelated unfortunate incident that disrupted the supply chain ecosystem was the Suez Canal blockage caused by the container ship Ever Given end of March 2021.[2] Globally operating cargo owners reported that they had limited visibility of their goods en route and the time of arrival at the different destinations post blockage. This incident

---

[1] Lind M., Lehmacher W., Hoffmann J., Jensen L., Notteboom T., Rydbergh T., Sand P., Haraldson S., White R., Becha H., Berglund P. (2021) Improving a congested maritime supply chain with time slot management for port calls, The Maritime Executive, June 29, 2021 (https://www.maritime-executive.com/editorials/how-time-slot-management-could-help-resolve-port-congestion)

[2] Lind M., Lehmacher W., Jensen L., Rydbergh T., Becha H., Rodriguez L. (2021) The Suez Canal puzzle– pulling the pieces together, The Maritime Executive, March 31, 2021 (https://www.maritime-executive.com/editorials/the-suez-canal-puzzle-pulling-the-pieces-together)

did not only create bottlenecks in ocean transport but also in air cargo between Asia and the United States of America and Europe and even train moves between Europe and Asia.

Today the parties involved, in particular, the cargo owners have limited options to intervene and hardly any chance to prioritize their cargo during such events. Many transport operations, for example, within the maritime sector, which carries more than 90 percent of global trade, are based on the principle of first come, first served. This means that operations are coordinated based on the physical presence rather than on the means of the digital economy, such as the principle of the appointment economy where activities are arranged in planned slots. This, together with limited supply chain visibility, means that the recipients lack sufficient situational awareness about the progress of the movement of shipments, leaving them unable to estimate and influence the time of the arrival of the goods. This also results in severe challenges in planning for succeeding operations, specifically, and the future level of required supply, in general.

Two changes in operating would make a big difference. First, extensive data sharing along the supply chain providing accurate localization information and visibility, improving forecast accuracy to enable the synchronization of the operations of the different carriers such as trucks, trains, barges, and ships carrying the goods. Second, the interoperable integration of systems empowered by digital data-sharing platforms that are home to powerful analytics, allowing to make sense out of the increasing amount of operational and ecosystem data to improve decision-making and automation in the longer run.[3]

The tide may have turned. We now observe that more and more physical objects are becoming digitally twinned with data fed into larger platforms.[4] Big data analytics and machine learning provide new opportunities to combine multiple data sources to improve decision-making. In that way, the invisible is made visible through data-informed reports on how different actors are actually performing in respect to their service level agreements. Also, on how potential disruptive events are emerging. Increased supply chain visibility allows decision-makers to take near real-time decisions based on increasing accuracy of information with the progress of the movements of the goods. The latter has been coined sequential analytics,[5] allowing to make changes of plans as late and as precisely as possible in the process. The closer we get to critical supply chain events, such as an airport visit or the final delivery of goods at an agreed location, the more accurate predictions are possible.

The volume of zettabytes of data generated that is growing at increasing speed allows for enhanced supply chain information and faster and fact-based decision-making. However, this requires that the different actors along the chains share their data, support common data architectures, and invest in building capabilities allowing for the proper selection and use of advanced digital tools and systems.

---

[3] Lind M., Becha H., Simha A., Bottin F., Larsen S.E. (2020) Digital Containerisation, Smart Maritime Network, June 18, 2020 (https://smartmaritimenetwork.com/wp-content/uploads/2020/06/Information-transparency-through-standardized-messaging-and-interfacing.pdf)

[4] Lind M., Becha H., Watson R.T., Kouwenhoven N., Zuesongdham P., Baldauf U. (2020) Digital twins for the maritime sector, Smart Maritime Network, July 15, 2020 (https://smartmaritimenetwork.com/wp-content/uploads/2020/07/Digital-twins-for-the-maritime-sector.pdf)

[5] Warren P. (2021) Reinforcement Learning and Stochastic Optimization: A unified framework for sequential decisions, Castle Labs, Princeton University (https://castlelab.princeton.edu/RLSO/)

Capacity building on global scale will take time. We also need a number of mindset shifts. The value of data does not emerge through holding onto it with the hope to be able to monetize one's own data in isolation and for the individual interests. The value of data unfolds once it is shared. This, particularly in form of efficiency gains through synchronization and integration across supply chain networks (see figure below). Data allows the various players to engage in digitally empowered collaboration to unlock hidden asset productivity (Figure 11.1).

FIGURE 11.1 Maritime Informatics enablers and expected effects. Lind M., Watson R.T., Lehmacher W. (2021) Key steps towards a high performing maritime industry, Container-News, March 27, 2021 (https://container-news.com/key-steps-to-a-high-performance-maritime-industry/) (www.maritimeinformatics.org).

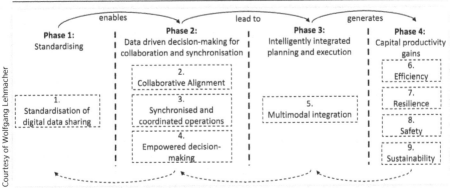

Standardized digital interfaces and collaborative alignment and processes are essential to allow for the integration of systems and augmented performance among the episodic coupled actors, for example, when a carrier visits a critical node. New knowledge is required to realize the vision. The digital transformation efforts start with raising awareness about benefits and opportunities. Yet about 80% of the ports in the world have not established or are not even planning to build the respective digital capabilities to allow for implementing digital architectures, systems, and applications that would allow them to synchronize port visits and optimize the utilization of infrastructure and resources along the goods value chains. Digitization of the goods mobility ecosystem requires an inclusive approach. If we fail in this respect, 80% of the maritime world will fall further behind.

On an International level, there are efforts made towards establishing digital performance indexes for ports and other types of transport nodes. Such indexes would both drive the digital inclusion for the nodes of the world and provide the foundation for a focused discourse and action plan on building capacity and digital capabilities. Sharing experiences among the different modes of transport would also inform the new thinking across the different goods transport systems.

How can we join forces to drive change across goods mobility systems and practices globally? There are many Wolfgangs and Mikaels out there in the industry that can support the larger goods mobility community in their efforts to raise their level of awareness and understanding around the new possibilities brought about by the digital economy. While the change will result from industry-wide knowledge exchange, as well as collaborative and open innovation, the breakthrough also needs common standards for interactions that are brought into use. Standardization bodies play an important role in the process. But what really counts is large-scale adoption of new technologies and the willingness to reengineer current practices in a way that they make optimal use of the new solutions and possibilities that are emerging. Initially, much of the effort needs to focus on instilling confidence and courage among business leaders and policy makers to pursue the necessary investments and initiatives to build capabilities and establish a truly digital environment as the basis for a new quality of economic growth in the interest of the economy and the common good. One major step towards this goal would be securing that coordination and synchronization within the supply chain is pursued without assets, like trucks and ships, being physically present.

Our aim is a global call for slot management at large[6] to be adopted for the different types of goods mobility systems to ensure the fluidity of the flows of goods. Unchoking the current goods mobility systems and their sustainable expansion requires a radical rethinking of current behaviors, practices, and models. We are convinced that a slot management approach across all nodes and modes allowing for near real-time status and progress updates would significantly increase the fluidity of the flows of goods and improve the use of resources and infrastructure deployed along the chains. We consider the realization of slot management practices as a major step forward for carriers, forwarders, and cargo owners to also raise their influence on supply chain decisions in their own and their customers' interest. Better synchronization and integration will also result in reduced carbon emissions, which would benefit our society and future generations. Building a fluid goods mobility system includes that we avoid paving the cow paths by introducing digitalization just as a means to cement existing inefficient patterns of operations rather than building better systems.[7]

About Wolfgang Lehmacher

Wolfgang Lehmacher is a global supply chain and technology strategist. He gained over 25 years of industry and leadership experience as an executive and consultant in organizations worldwide. He is advising and actively supporting corporates, asset owners, international organizations, governments, and startups. Wolfgang was Director of Supply Chain and Transport Industries at the World Economic Forum in New York and Geneva, Partner and Managing Director for China and India at CVA in Hong Kong, President and CEO at GeoPost Intercontinental in Paris, and

---

[6] Lind M., Lehmacher W., Jensen L., Notteboom T., Rydbergh T., White R., Becha H., Rodriguez L., Sand P. (2021) Resolving the ship backlog puzzle in the Suez Canal: Predicting ship transits in capacity-constrained areas, The Smart Maritime Network, April 22, 2021 (https://smartmaritimenetwork.com/2021/04/22/resolving-the-suez-backlog-predicting-ship-transits-in-capacity-constrained-areas/)

[7] Lind M., Becha H., Simha A., Larsen S.E., Ben-Amram E., Marchand D. (2020) The maritime ecosystem needs ecosystem innovation to avoid "paving the cow paths", The Maritime Executive, December 12, 2020 (https://www.maritime-executive.com/editorials/maritime-ecosystem-needs-innovation-to-avoid-paving-the-cow)

Head of Eastern European and Eastern Mediterranean Regions, and Country General Manager Switzerland, at TNT. He is the advisory board member of The Logistics and Supply Chain Management Society in Singapore and Ambassador of the European Freight and Logistics Leaders Forum. He is a Founding Member of the Logistikweisen, a logistics expert committee under the patronage of the German Federal Ministry BMVI, and NEXST, a think tank initiated by Reefknot Investments, Kuehne + Nagel, and SGInnovate in Singapore.

About Mikael Lind

Dr Mikael Lind is (Adjunct) Professor in Maritime Informatics at Chalmers University of Technology and Senior Strategic Research Advisor at RISE. He has initiated and headed a substantial part of several open innovation initiatives related to ICT for sustainable transports of people and goods. In his capacity as the world's first professor in Maritime Informatics, he is also part-time at the Chalmers University of Technology (M2), Sweden, exploring the opportunity of maritime informatics as an applied research field. Lind serves as an expert for World Economic Forum, Europe's Digital Transport Logistic Forum (DTLF), and UN/CEFACT. He has been the lead author of many concept notes associated with maritime and transport informatics brought up by the international trade press and has become a recognized thought leader in Maritime Informatics. Lind has also served as mini-track chair for Maritime Informatics at the major regional IS conferences in Europe and the Americas for several years.

Shoshana Lew

Executive Director
Colorado Department of Transportation
(CDOT)

When I became Executive Director of the Colorado Department of Transportation (CDOT) in 2018, I knew that we had to plan for a future with reduced emissions but also change the way we thought about planning itself.

In May of 2019, our team at the department embarked on an effort to refresh our transportation plan and priorities based on firsthand input from residents across the state. Our goals were simple: to hear directly from Coloradans about what they needed from our state transportation system, to ensure that we're prioritizing precious taxpayer dollars in ways that best deliver on those needs, and to energize an ongoing statewide conversation about the vitality of transportation in connecting our daily lives.

The months that followed were an exciting journey for us as we visited every county in Colorado, meeting with community leaders and residents in their own towns. For me, a newcomer to Colorado at the time, the tour was a way to "take a walk in your shoes" and experience what Coloradans experience on a daily basis as they travel around their communities and across the state. This included personally attending many county, regional, and stakeholder discussions. It meant rush-hour drives up and down I-25, seeing stunning vistas along the rural roads of Hinsdale and Mineral counties, engaging in discussions about freight at the Lincoln County Fair, having conversations about defense access needs in the Pikes Peak area, and visiting communities across the east-west I-70 Mountain Corridor to the Western Slope and down to the Four Corners in Colorado's farthest southwest point.

Throughout these conversations, I was struck both by the uniqueness of each community and by the common themes that emerged when talking about the challenges, frustrations, hopes, and aspirations that transportation infrastructure evokes. From residents along the Front Range (the north-south region that runs along I-25 parallel to the Rockies and includes Denver, Boulder, Colorado Springs, and Pueblo), it was no surprise that we heard a lot about the pressures that come with rapid growth, not just the traffic, but the broader uncertainty that accompanies population density and a built environment that is changing rapidly—a sense that small towns can become big cities seemingly overnight.

Keeping up with the associated transportation demands requires dollar figures that can seem daunting and a vision for how to efficiently and sustainably connect people and economies in ways that preserve the quality of place and reduce road congestion and air pollution. Currently, the transportation sector—everything from cars to freight to airplanes—is the leading source of emissions in our state, as it already is nationwide.

That is why, in tandem with implementing CDOT's 10-year plan, executive agencies partnered with the state legislature to develop and enact historic legislation to modernize Colorado's transportation revenue, bolster our economy as it emerges from COVID-19, and tackle the interface between transportation and climate change.

That legislation, Senate Bill 260, was signed into law by Gov. Jared Polis on June 17, 2021 and combines a revenue structure that includes up-front stimulus with road usage charges that address the emerging changes in how we utilize transportation—such as ride sharing, electric vehicles, and retail delivery—as well as charges that accrue to drivers of conventional vehicles who use the roads.

The revenue from the over $5 billion package pays for a mix of projects in the state's 10-year plan with investments in local roads, accelerating electrification, making investments in climate mitigation, and expanding the state's successful program to improve main streets across Colorado.

But dollars alone can't determine the future of our transportation system—and SB260 also includes significant provisions to address planning, account for pollution impacts, advance equity, and improve transparency and accountability. In short, we have to fundamentally rethink the way we use land, plan transportation projects, and get people and goods to their destinations. Again, transportation is the leading source of emissions in Colorado and nationally.

In January of 2021, the CDOT began a process of meeting with hundreds of stakeholders around the state to discuss a plan to lay down a new set of standards for transportation planning in an effort to reduce greenhouse gas emissions and meet our state's climate goals in the future. How we build projects determines driving levels, and that, in turn, determines emission levels.

Our proposal is now moving forward in an official rulemaking process, with public hearings around the state. Upon adoption of this rule, Colorado would become only the second state in the nation to establish greenhouse gas-related reduction requirements on transportation planning—and the first in the Intermountain West.

The proposed rule would require our department and the state's five metropolitan planning organizations (MPOs) to determine the total pollution and greenhouse gas emission increase or decrease expected from future transportation projects and take steps to ensure that greenhouse gas emission levels do not exceed set reduction amounts. This approach will also streamline the planning and delivery of innovations that have proven successful in improving quality of life and air quality, like adding sidewalks, improving downtowns for active transportation with "complete streets," improving local and intercity transit and first-and-last-mile connectivity to transit facilities, and adding bike-shares.

If adopted, the greenhouse gas reduction levels proposed in this rule for 2030 are equivalent to burning 169 million fewer gallons of gasoline or taking approximately 300,000 cars off the road for a year.

The proposed rule requires the department and metropolitan planning organizations to achieve set greenhouse gas reduction levels at four different time periods (2025, 2030, 2040, 2050). If they can't demonstrate these reductions, even after employing other transportation strategies to reduce emissions, the Colorado Transportation Commission must restrict the use of certain funds—essentially requiring dollars be focused on projects that help reduce transportation emissions. The draft rule includes provisions for the commission to waive these funding restrictions under certain limited circumstances.

We've also made some changes internally at the CDOT that are aimed at aligning our organization with a greenhouse gas-reduced future. In 2019, we established the Office of Innovative Mobility, which integrates the department's multimodal efforts through the Division of Transit and Rail with an emerging focus on incorporating electrification and other zero-emissions vehicles into our system and equipping our infrastructure to accommodate them. This office reports directly to me, and it has recruited new leadership to elevate the role of the Division of Transit and Rail, along with expanding expertise on electrification and mobility choices.

The department is also closely involved in an effort to develop Front Range Passenger Rail, a 180-mile train line that would connect people along the I-25 corridor, which now sees heavy congestion. The Front Range is also where 85 percent of Coloradans live.

We know we have to give people other options besides driving, and department engineers and planners have been doing the early analysis to get this project—which Amtrak has identified as one of its top new corridors for expansion—to the next stage.

Front Range Passenger Rail would work in tandem with the department's existing Bustang intercity bus service, which has excellent ridership. We have been expanding Bustang for the last several years since it first began, and now it serves most every corner of the state. The department has long-term plans to link Bustang with other services with strategically placed mobility hubs across the state.

All of these efforts—including the department's programs to spur micromobility through more bike lanes and to encourage employers to allow their workers to telecommute—are aimed at creating a future that is lower in emissions, a future that is sustainable, and a future that keeps Colorado beautiful and wild.

About Shoshana Lew

Shoshana Lew is the executive director for the CDOT. She is charged with leading the department in planning for and addressing Colorado's transportation needs and overseeing 3,000 employees statewide and an annual budget this year of approximately $2 billion.

Prior to coming to Colorado, she served as Chief Operating Officer (COO) for the Rhode Island Department of Transportation (RIDOT). Also she was Chief Financial Officer and Assistant Secretary for Budget and Programs for the US Department of Transportation (USDOT) as well as Deputy Assistant Secretary for Transportation Policy at USDOT. At USDOT, Lew also managed departmental efforts to accelerate permitting and delivery of infrastructure projects, development and implementation of fuel economy standards for cars and trucks, and departmental coordination on other key regulatory matters, including asset management for highway and transit programs.

Lew has also worked in other areas of the federal government, including the US Department of the Interior and the Office of Management and Budget, where she focused on the implementation of the American Recovery and Reinvestment Act, and at the White House, where she served as Senior Adviser at the US Department of the Interior's Bureau of Ocean Energy Management and as a policy adviser at the White House Domestic Policy Council, where she focused on energy policy.

Prior to her federal service, she worked at the Brookings Institution. She is a graduate of Harvard College with a Bachelor of Arts in History, Ms. Lew also has an Master of Arts in History from Northwestern University.

Courtesy of Alisyn Malek

Alisyn Malek

Executive Director
Coalition for Reimagined Mobility

I started my career in transportation and mobility because I wanted to positively impact climate change. In 2008, I graduated from the University of Michigan with a degree in Mechanical Engineering and took an internship with General Motors (GM), where I worked on the Chevrolet Volt. I soon became intrigued by the possibility of working on clean transportation—electric vehicles, specifically—for one of the world's largest automakers. As my internship came to an end, GM brought me on full time, officially launching my career.

When I started, I was one of only three people in the company working on electric vehicle charging. This offered me more latitude and exposure than is typical for an automotive engineer fresh out of college. I was fortunate to participate in GM's work with other industries as they learned about this new technology and tried to understand how it would evolve. In addition to working with suppliers on vehicle components, I contributed to standards committees where I learned about the complex negotiations that take place among industry participants to commonize something as seemingly simple as the design of a charger plug.

The work was new and exciting, and sparked curiosity and deep conversation. For instance, how would utility companies supply power for the uptick in charging? How could we equip dealerships to embrace the new technology, and how could we train electricians to work on high-voltage batteries and electric motors? Would people charge their vehicle at home or at work? If at home, what would we do for people who have to park on the street? As we explored each scenario, I came to appreciate the extent to which our world is built for gasoline-powered vehicles and the significant paradigm shift that must happen across a variety of industries to change that model.

One day during my time working on hybrid and electric vehicles, I received an upsetting phone call that altered my view on what mobility means for our society: my grandmother fell down the basement stairs and broke her hip. A few years prior, she had moved from the home she had lived in for five decades, with the intention of downsizing and living closer to her doctors and family. Now, as she slowly recovered from her injury, she could no longer drive. What's more, because her community did not have good alternate forms of transportation, she became dependent on others to drive her places and run errands for her. By moving her to a larger town, we thought we were improving her quality of life, though did not consider how car dependent we were making her until it was too late.

In contrast to my grandmother's situation is my grandmother-in-law's situation. A year younger than my grandmother, she still lives in the same house that she and her husband bought over half a century ago. It's in a suburb of Detroit, with a Knights of Columbus Hall and a bowling league nearby. After a shoulder surgery limited her ability to drive, she begrudgingly accepted that she had access to a bus that she could call and arrange for rides to her doctor's appointments, in addition to relying on friends for rides to luncheons and other social events. It has been an adjustment for her, and it's not perfect, though she lives more independently and socially than my own grandmother. Her location affords her access to more mobility options and supports her in leveraging her social network.

During the evolution of these women's lives, I have worked on electric vehicles, autonomous vehicles, venture investing, and policy. It has been a joy to innovate at that cutting edge of technology with the ability to glimpse at the future. What I see missing, though, is the critical perspective on the idea that more than just car technology needs to change: the entire system of how we and our goods get around needs to change as well, and that requires action. I want to be independent like my grandmother-in-law when I'm in my 80s. However, I don't want to have to call for a ride 24 hours in advance in order to catch the bus, like she does. To help create a higher and healthier quality of life for everyone, we need better options to keep people rooted in their communities and social networks. Transportation and the systems that support mobility play a huge role in how we make that happen.

Potential technologies like autonomous vehicles and existing services like Amazon's ordering and fast delivery could help solve these challenges, while at the same time helping to reduce carbon emissions However, these technologies are usually designed for able-bodied, tech-savvy people who are often wealthier than the average American. With 10 percent of elderly adults in the United States living under the poverty line,[1] and 35 percent of them having some form of disability,[2] this technology is not necessarily targeted at solving—or even taking into consideration— their problems.

These realities form my view on the future of mobility. I want a future where I can enjoy equal access to goods and services without impacting the global climate. I want this for my parents' generation, too. This means easier access to basics like doctors' appointments through telehealth or local clinics, or delivery from grocery

---

[1] https://www.kff.org/medicare/issue-brief/how-many-seniors-live-in-poverty/
[2] https://disabilitycompendium.org/sites/default/files/user-uploads/2017_AnnualReport_2017_FINAL.pdf

stores and small neighborhood shops. These types of changes require technology, of course, but also a rethinking of land use in the US. We need to rethink everything from the layout of residential areas to the logistics of shipping goods directly to homes instead of to large box stores. A lot of change is required.

Transportation as we know it today—such as roads and cars, and their maintenance—is really expensive. Through better planning, utilization, and the application of technology, we have the opportunity to focus on identifying the true sources of value in our transportation systems while eliminating the wasted costs. Industry leaders will need to envision a new orchestration of technology, connectivity, shipping, road and sidewalk design, and land use. This will require the first major rethink of the entire transportation system since the mass introduction of the automobile in the 1930s. I believe that by doing so we can truly develop travel options that align more closely with our individual values and needs.

About Alisyn Malek

Alisyn Malek is a change maker in the mobility and automotive sectors. With over 12 years of experience leading organizations, driving investments, and developing products, Alisyn has played an active role in tackling complex challenges head on and offering effective solutions.

Currently, Alisyn is Executive Director for the Coalition for Reimagined Mobility, whose purpose is to help inform, streamline, and modernize policies that address the changing global mobility landscape we face today through an interconnected, systemic approach that is rooted in research. The ultimate goal is to enable more inclusive transportation systems that leverage technology while ensuring a people-first approach to enacting public policy. Additionally, she is the founder and CEO of Middle Third, a boutique mobility consultancy helping clients to unlock the potential of technology and transportation.

Prior to this role, her unparalleled understanding of how automotive and innovative technologies work together inspired her to co-found May Mobility, building an autonomous vehicle transportation solution that would solve urban transportation challenges ahead of the competition. As COO, she grew the company from laboratory concept to operations in three states and a strong pipeline for growth in less than three years creating the finance, marketing, sales, customer support, and field operations teams. Alisyn started her career as an engineer working on electric vehicle charging technology at GM before moving to their corporate venture arm, leading investments across electrification, connected vehicles, mobility, and autonomy. During that time, one of the most notable investments Alisyn led was the acquisition of Cruise Automation to further accelerate GM's development of autonomous vehicle technology.

Alisyn was recognized as an *Automotive News* All Star in 2019, a top ten female innovator to watch by *Smithsonian* in 2018, and named a top automotive professional under 35 to watch by LinkedIn for her work in cutting-edge product development and corporate venture.

Susanne Murtha

National Lead for Connected and Automated Technology
AECOM

Safety. I've been steeped in safety culture and leadership as a Girl Scout and Boy Scout leader for over twenty years. I've always worked for organizations with a safety focus. Safety is the most basic, fundamental transportation problem that impacts all of us—drivers, riders, pedestrians, and other vulnerable road users. It's a huge amount of 40,000 fatalities that we can stop. What if the efforts I lead and support could contribute to saving 40,000 people in the US and 1.3 million throughout the world?

No one can argue with safety. On September 11, 2001, I was working as an economist in Boston. In the days following the attack, I watched the bombers leave Hanscom Air Force Base heading East over the Atlantic. I lost friends in the attacks on New York. I tried to enlist, but asthma prevented me from being accepted, so I bought a house a mile from the Pentagon and moved to Washington DC to try to use my specific experience in transportation to improve safety for people in a civilian role. I ran the first ITS (Intelligent Transportation Society) America practice linking vehicles and infrastructures. I supported the standardization of V2X (vehicle-to-everything connection) communications with the OEMs, the state DOTs, and the federal government. I helped write the FCC rules for use of the 5.9 GHz spectrum.

Since then, I've worked with several other firms, all on the trajectory to deploy safety technology for transportation. Not everything worked as planned. Not everything will work as we're planning now, but it's all better than 40,000 fatalities. There are a lot of important reasons to improve how we do transportation. No one can argue with our fundamental need for safety, and short of changing peoples' behavior, we can use technology to save lives.

Decades of millions of global fatalities, showing no substantial reduction (except after seatbelt mandates) tells us we need a sea change in our approach to transportation. Human behavior can't be changed, we've seen that. The solution, then, is to remove the human from the tasks of transportation, which are causing fatalities. Automation is defined as a machine doing the tasks of humans. Many transportation professionals use the term "autonomous" or "autonomy" when describing vehicles, which have aspects of machine operation. I am careful to use the term "automation" as this term most closely describes the shift of functions of the driving task from a person to a machine, which is the fundamental cause of these millions of fatalities.

"Autonomous" means when something functions independently of other things or people. No aspects of transportation are autonomous. Transportation is multimodal and interdependent with the infrastructure, road users, and nonautomated vehicles. Solutions for improving transport must be broad and multimodal, and that is the current focus of the work of my team. We're working with infrastructure owners and operators (IOOs) such as DOTs, tolling authorities, cities, counties, transit authorities, and airports to plan fully automated trips. We're collaborating across many practice areas to bring many of our clients together to create automated trips. We're supporting the deployment of eVTOLs (electric vertical take off and landing aircraft), highly automated transit buses, commercial vehicles, and light vehicles.

Our teams are using automation systems to remove human error from transportation. Approximately 95% of crashes are because of human error. When we create systems that take the most challenging driving tasks from the driver, we can finally start to close in on reducing that staggering number of fatalities. More good news—in addition to safety—there are several beneficial outcomes to deploying automated technologies. An automated transportation system is more efficient. Most automated vehicles are electric and will have a net positive impact on the environment. Also, fewer crashes mean faster throughput of vehicles, which means less time spent in traffic and lower emissions from vehicles. Safety is the cornerstone. If we can get safety right with automation, we'll be able to fundamentally shift how people and goods move. We will save over a million people each year.

About Suzanne Murtha

Suzanne Murtha is the National Lead for Connected and Automated Technology at AECOM. She oversees the growth of the practice related to connected and automated vehicles, as well as the surface integration of eVTOLs and advanced air mobility. Her work accomplishments span advanced technology growth in intelligent transportation, tolling, and, most recently, the leadership of the practice of urban air mobility and automation of full trips for goods and people.

Suzanne is on the Board of Directors of the OmniAir Consortium and AUVSI and is the Chair for the ITS America eVTOL Task Force and the AUVSI (Association for Unmanned Vehicle Systems International) AV Goods Movement Task Force, and leads the V2X RUC and Tolling Certification Task Force for OmniAir. She is a 25-year veteran of the ITS industry and is the recipient of the 2021 AUVSI Member of the Year Award for her work in the automation of freight movement. Suzanne is also the recipient of the Girl Scout Council of the Nation's Capital Outstanding Leadership Award for over 20 years of volunteering in scouts. She is based in Alexandria, Virginia, and Punta Gorda, FL.

13

Mary Nichols

Former Chair
California Air Resources Board,
2007 to 2020 and 1979 to 1983

Courtesy of Mary Nichols

I bought my oldest grandchild her first car when she was a little over one year old. It was a powder blue plastic one-seater that she could sit in and push with her feet or be pushed with a detachable stick, while turning the steering wheel. We both got a lot of joy taking it for spins around the block, pointing out squirrels and flowers. That first taste of mobility beyond her own home brought opportunities for connection, learning, growth. And while I am not promoting the toy car as the way to solve the pressing environmental and social equity problems embedded in our current transportation system, I do think it is important to acknowledge the power of transportation to improve lives.

Looking even farther back, when I first moved to Los Angeles 50 years ago, you couldn't see the San Gabriel Mountains most days because of the smog that smothered the city. Those days are now the exception, not the rule. Thanks to public demand, political will, and industry creativity, vehicles today emit less than 1% of the pollution they did before 1970. But economic and population growth, the concentration of low-income residents along freight movement corridors and congested roadways, and a lack of public investment in transit options have left too many people exposed to unhealthy air pollution.

When we look beyond our own borders, we see cities in the developing world living with worse air pollution than Los Angeles in the 1950s, as well as increasing emissions of greenhouse gases that are causing global warming to advance faster than the scientists' models had predicted, with up to half of that coming from growing car and truck traffic. Zero-emission vehicles are an exciting technology that with every passing day become more and more part of the present rather than some off-in-the-distant future. We have to keep our foot on the pedal to keep these options flowing faster and more broadly. It's not just about the cars though. We also need to re-envision how we design and reshape our cities and communities.

I began my career as a public interest environmental attorney in California in 1972. At the time Southern California was just beginning to grapple with the impacts of unchecked pollution from the cars that a growing population craved for the freedom they provided people to move around wherever they wanted whenever they wanted. The Los Angeles metropolitan region spread out from the coast to the mountains that ring the basin, replacing orchards and farms with tract houses and low-rise commercial development. As the population grew and the economy grew even faster, smog forced children to stay indoors during school recess. The solution advanced by transportation planners: build 8 and 10 lane freeways in a 2-mile grid pattern across the region to speed the cars and trucks more quickly through the area. Brand-new environmental statutes passed just after Earth Day brought this and the related consequences of unchecked transportation growth to national prominence, my public interest law group sued to stop a proposed freeway that would have pushed an already growing pollution problem even further into the red zone. The freeway ultimately got built. But not before the courts forced the State to add a new mass transit line and replace many hundreds of houses that had been demolished to make way for the project. It was better, and it marked the end of an era of major new highways, but still did not make a dent in the region's commitment to single-occupancy vehicle transportation. What allowed that lifestyle to remain viable was tremendous technological change.

Cars today are cleaner than anyone could have imagined at that time. Over 50 years of government regulation demanding improvements in emissions and fuel economy produced innovations in technology, better manufacturing techniques, and cleaner fuels. We can document the lives saved and serious illnesses prevented as a result of these changes. Yet transportation still accounts for around half of the dangerous air pollution and climate-forcing emissions in the world. These impacts are often most pronounced in the places that can least afford them—like communities that live near mega-warehouses, goods movement corridors, and webs of tangled freeways. At the same time, there are too many people across the planet that still don't have access to the safe, affordable, and sustainable mobility options that they need.

What makes me optimistic that we will create a future of mobility that is better aligned with our needs is the unprecedented amount of creativity now being directed to a comprehensive array of solutions. Zero-emission vehicles are here, manufacturers are offering more and different designs, costs are getting closer to parity with combustion versions, and have already shown that they provide a superior driving experience. New micromobility choices are popping up, and new models for shared-use vehicles are appearing in cities. Automation technologies open a huge range of new, cleaner, more efficient ways to move people and stuff around. We have a whole set of emerging data and technology tools that we can use to start giving more people more mobility choices and make goods movement more efficient. People are demanding jobs and housing that don't demand that they devote hours of each week to sitting behind a steering wheel. Communities coping with the economic devastation of COVID-19 learned to quickly transform outdoor parking spaces and concrete zones on many of our streets into welcoming, useable places.

Creating this lively, adaptable mobility future demands a combination of vision, informed rules and regulations, and popular demand that creates a market incentive to bring newer, better, clean, safer, and more sustainable mobility options to the market in ways that are accessible to everyone. It's already happening.

About Mary Nichols

Mary Nichols served as Chair of the California Air Resources Board from 2007 to 2020, a post she also held from 1979 to 1983. As the longest-serving head of the nation's premier clean air agency, appointed by three different governors, Nichols is credited with pioneering innovative regulatory and market-based programs that have improved public health and the environment across the country and around the world. Throughout a career devoted to public and not-for-profit service, she has been recognized for a commitment to evidence-based problem solving that is inclusive of all stakeholders and focused on providing benefits to all communities, particularly those most impacted by pollution and other health threats.

She currently is Distinguished Visiting Fellow in the Columbia Center on Global Energy Policy, Senior Visiting Fellow at the Cornell Atkinson Center for Sustainability, Professor of Law at UCLA Law School, and Vice Chair of the California-China Climate Institute at the University of California.

Courtesy of Trevor Pawl

Trevor Pawl

Chief Mobility Officer
State of Michigan
Office of Future Mobility and Electrification

Detroit in 2072: Putting a Face on the Future of Cities and Mobility

Imagine we are 50 years into the future. And you are living in Detroit, Michigan. Walking down the street, there are two things you immediately notice have changed dramatically since 2022. How Detroit looks and how its people move.

How Detroit Looks

The city does not resemble a post-apocalyptic Robocop movie set. In fact, the economic comeback that started in the early part of the twenty-first century never slowed. The city has swelled from less than 700,000 people in 2022 to 3 million in 2072. And Detroit is global. Ten times as many Detroiters now identify themselves as multiracial compared to a half-century earlier.

Detroit and other Great Lakes cities like Cleveland and Buffalo, known for being loops holding up America's 20th Century "Rust Belt" are booming. Engineers and designers from around the world have flocked to the city because of its reputation as America's hotbed for jobs in artificial intelligence, robotics, and the future of transportation.

As population density has grown, so has Detroit's skyline. The historic Renaissance Center now looks up at seven other buildings. Also, Detroit's skyline officially has the world's largest surviving collection of early twentieth-century gilded and art deco buildings thanks to early twenty-first-century activists.

But Detroit's new building boom of the twenty-first century hasn't been focused on industrial as it might in the past. Instead, it's focused on sustainability.

Detroit is a 100% carbon-neutral city with an energy mix that includes the grid, thermonuclear plants, biofuels, wind turbines, garbage automatically sorted via nanotubes, fog catchers, solar energy stored in glass and asphalt, and geothermal energy stored in molten salt (mined from the 100 miles of salt mines located directly under the city and its suburbs).

Energy plants will have changed a lot in 50 years. In the 2050s, power companies built two trash-burning powerplants off Jefferson and Michigan Avenues. Each plant has a 2,000-foot ski slope and hiking trails on its roof. Instead of blowing smoke, these plants blow steam rings from the over 800,000 tons of trash used each year to generate power and water for 150,000 homes and offices.

Despite its upfront costs, Detroit's commitment to a zero-carbon footprint has made the city rich. For example, repurposed carbon and landfill waste is now a popular building material. This has literally turned Detroit's trash into cash.

Much like how telephone lines went away in the 2020s, water pipes have also become a thing of the past. New technology introduced in the 2040s allows several Detroit buildings to create their own water supply through atmospheric water generators. These machines literally pull water out of the sky.

And Detroit's skyscrapers are no longer just for people. Fourteen-story vertical indoor farms dot the skyline. Growing cash crops using a fraction of the land and water taken up by traditional farms, without soil or chemicals. Wealth promotes health.

How Detroit Moves

The second thing that will have dramatically changed since 2022 is how Detroit moves. Believe it or not, Metropolitan Detroit and other metro areas are not driverless as many people thought earlier in the century. People love driving. Even if it's not the most efficient option.

Many of Detroit's suburban streets still allow self-driving vehicles and drivers to share the road. But at slower speeds, with drivers having only one lane. By law, all moving vehicles must be connected to the surrounding traffic infrastructure. This means even classic cars, like the "Imported from Detroit" 2012 Chrysler 300, must be retrofitted with new technology.

Truth is, in the year 2072, it's expensive and inefficient to own and drive a vehicle. There are heavy taxes, empty seat fees, and less parking spots.

This makes autonomous vehicles the primary mode of transportation for most Detroiters. And the fastest way to get around Metro Detroit is through mobility subscription plans. These plans give everyone access to a variety of electrified options: autonomous vehicles, shuttles, scooters, e-rickshaws, and e-bikes. Fleet management companies like Ford, GM, Uber, Lyft, Delta Airlines, and Amazon sell these plans. Each offering mobility on a per ride and per month basis.

Through the years starting in the 2020s, local and national leaders realized that the best way to fight congestion, reduce accidents, and uncover new city revenue was to redesign city streets to serve people first and vehicles second. This trend made it to every major and mid-sized city by the late 2050s. Putting people (and not vehicles) first meant providing more transport options and moving traffic below or above ground.

For example, in Detroit, Jefferson Avenue's ten lanes of traffic were moved underground into well-lit tunnels with no stop signs or red lights, just signs that give your estimated time of arrival. And above those tunnels, streams and waterfalls from clean water runoff were elevated and exposed next to walkways and green space.

A little further north in Detroit's Midtown neighborhood, Cass Avenue reimagined what an avenue could look like without lanes. Battery electric vehicles and citizens operate in harmony, aware of each other thanks to advanced artificial intelligence and 24/7 real-time digital mapping. Streets like Cass are a lot quieter now. No need for car horns and sirens, sensors pick up close calls or emergencies. And charging stations are hidden (embedded in the pavement).

Let's talk about autonomous vehicles themselves, which operate on 12-hour shifts before returning to nests along Metro Detroit's interstates. While some still have a front seat and backseat, many have designs that let riders interact, order lunch, entertain children, or even workout.

And although some vehicle interiors are sleek with modern curves, many resemble the backseat comfort of a late 1970s Cadillac. Upon arrival, your vehicle syncs with your 3D-hologram phone without breaching personal data. It also senses mood preferences through wearable technology.

The future of wearables has evolved to disappear into personal items like wedding rings and earrings. They use body heat and solar to charge. Wearables interacting with vehicle sensors can tell if a rider is in danger, requires medical attention, or simply prefers the passenger window to display images of a sunny day during five straight days of rain.

One part of the city that has transformed the most in 50 years, thanks to advancements in mobility, is Detroit's Eastern Market, which now operates every day. To keep pace, vendors don't require cash or card anymore. Sensors on your wearables sync with secured payment systems.

Under the market, autonomous delivery trucks only inches apart move like a swarm of bees transporting goods and letting droids and drones handle the doorstep deliveries. And above the market, super bikeways shoot e-bikes into and out of the city.

Detroit has become a leader in urban air transit, or "flying cars." These small, electric aircraft take-off and land vertically all over Metro Detroit. Turning a two-hour commute from Detroit to Ann Arbor into 10-minute flight. Imagine taking the elevator up to go home from work because Uber is waiting on the roof. Just as skyscrapers let cities leverage land more efficiently. Urban aircrafts let cities leverage roads more efficiently.

Repurposed highway bridges now operate as "air hubs" for landing and charging these aircrafts. And, besides having one of the world's largest regional fleets, Detroit is a world leader in designing and manufacturing these aircrafts because of its assembly line capacity and centuries of engineering expertise.

For longer flights, Detroit Metropolitan Wayne County Airport (DTW) is no longer only an airport. DTW is also a major stop on the Americas Hyperloop line and the World Hyperloop line (connecting New York to London via Russia and Alaska). These levitating Hyperloop trains glide at airline speed through a network of vacuum tubes. Hyperloop trains can get from Detroit to Chicago in 30 minutes.

All of these transportation options make Detroit not just the center of the global mobility industry in 2072, but also one of the best cities in the world to simply get around.

The best news? Every concept and technology that has been mentioned is already in development somewhere. With bold, clever action and a 50-year head start, why couldn't Detroiters bring this vision to life? After all, it would still be easier than their first act, which was putting the world on wheels. But for the Motor City, or any other city, the time to start is now.

About Trevor Pawl

Trevor Pawl is Chief Mobility Officer for the State of Michigan and leads Michigan's Office of Future Mobility and Electrification. In this position, Pawl is responsible for working across state government, academia, and private industry to grow Michigan's mobility ecosystem through strategic policy recommendations and new support services for companies focused on the future of transportation.

Prior to this position, Pawl served as Senior Vice President of Business Innovation at the Michigan Economic Development Corporation, where he led the state's programs for mobility, supply chain development, export programs, and entrepreneurial assistance.

Pawl has been named *Crain's Detroit Business*'s "40 Under 40" and "50 Names to Know in Government." He's also been named Development Counsellors International's "40 Under 40 Rising Stars of Economic Development" and the Great Lakes Women's Business Council's "Government Advocate of the Year."

Pawl holds a bachelor's degree in Economics and Marketing from Grand Valley State University and a Master of Business Administration from the University of Detroit Mercy where he currently serves as an adjunct professor.

John Peracchio

Strategic Consultant
Intelligent Transportation Systems
("ITS") Sector and Automotive Industry

To begin, I admit that I am decidedly not a futurist. Indeed, I have a tough time dealing with the here and now along with the near-to-intermediate term for my consulting clients. And so, I shall confine my musings to what I see happening on the short end of the given timeline: 20 years from now. My overarching message is that the promise of technology is significantly overstated and that our collective will may not coalesce to overcome mobility challenges, especially involving electrification.

A personal story involving my vehicle, at this writing a late-model compact SUV made by a prominent German automobile company, illustrates the concern. One day I walked into my garage to find all of the windows and the sunroof open, apparently happening overnight. This went on for weeks. I woke up early one morning and while I am in the garage, the vehicle flashed its headlights and opened the windows/sunroof by itself ... in my very presence. Alarmed but undaunted, I got the key fob, turned the vehicle on, and as I attempted to close the windows and sunroof, the vehicle defied me and lowered the windows and opened the sunroof the second I took my finger off the switches! Ten days before I was to bring the vehicle in to the dealer for the service appointment I made, it ceased its demonic ways. The dealer's technician could find no error code to explain the vehicle's autonomous behavior. Happily, in 20 years, I predict this problem will be fixed.

Let's consider the easy stuff. There will not be a ubiquitous deployment of fully automated vehicles. I use the term "fully automated" versus "fully autonomous" as the latter connotes a robot deciding where to go where the former connotes a human in charge. I do see fully automated vehicles being deployed in carefully geofenced areas in which hyper-accurate/high-definition maps are available to augment the vehicle's onboard sensors. Microtransit and long-haul heavy truck routes leap to mind here. Also these contraptions will be segregated from mixed traffic with physical barriers to avoid inevitable collisions with human-controlled vehicles. It turns out to be difficult for humans to understand what a robot car will do next (robotic intent), which is why many of the collisions involving fully automated test vehicles are rear-enders.

The hardware and software limitations of sensors, edge processing, and artificial intelligence in terms of replacing the human brain will remain myriad. Having a million lines of code in a vehicle is a dubious distinction. It just means a million things can go wrong. Machine learning is very dependent on the data being supplied. Bad data leads to bad outcomes. Try this at home. Go to websites that involve a pastime you don't really like, say fishing. Very soon, every time you log onto the internet, you will see ads for fishing rods, boats, and clothing. There is a cybersecurity aspect to this as well. Imagine a nefarious actor feeding your car corrupt data.

All is not lost, however. Technology and innovation attending the development of fully automated vehicles will definitely make human driving safer and offer a better driver experience. Advanced driver assistance systems will save many more lives, but the automobile industry will never shake challenges with software, which has become an increasing reason for recalls of late. Who hasn't experienced some trepidation when your computer's firmware has been updated overnight?

In 20 years, I believe electric vehicles will become the majority of *new* vehicle sales at least in North America, but only by a slim percentage. And the car park at that point in time will remain overwhelmingly dominated by internal combustion vehicles. There are many reasons for this including customer hesitation driven by a variety of factors and the simple fact that vehicles last longer. The biggest challenge facing volume EV deployment, however, is the lack of charging infrastructure and the inadequacies and vulnerabilities of the electric power grid.

Wild predictions from politicians, pundits, and industry "experts" aside, there are too many obstacles in the way of building out charging stations in volume combined with the fragility of our electricity distribution system for EVs to become a dominant form of personal transportation any time soon. Massive investment in the top-level grid is certainly required, but just as important is the need to enhance or replace local electricity distribution; think of the transformers on the street where you live, many of which are old and lacking capacity for handling the load of charging EVs at any time of day or night.

Of course, utilities will be pestering public service commissions to obtain "cost recovery" to fix all of this with rate increases, but they will have mixed success as politicians struggle with competing policy considerations. Climate change is real and serious, but should all ratepayers pay for charging infrastructure benefiting EV owners? Or is this what everyone must pay to reduce greenhouse gas emissions? And what about consumer incentives to purchase EVs? My own view is that we will need them, but they should be calibrated to individual income levels. I am not much interested in subsidizing with tax dollars the purchase of an EV by my rich friends.

On the plus side, the cost of batteries will have plummeted in 20 years, making EVs affordable to most consumers. And there will be far fewer recalls involving EVs that spontaneously combust. Perhaps most significantly, regional and local units of government will have modified building codes to make new, single-family homes and multi-unit dwellings "EV ready", and there will be many more public fast-charging stations available.

By the way, alternative power generation (e.g., solar and wind), coupled with increased electricity storage capacity, will greatly support EV deployment, especially in remote locations where running power is otherwise prohibitively expensive. To this end, we will necessarily question the monopoly granted to public utilities. The legal and regulatory framework that served us well in the past will require modification to get us where we want to be in the future.

Now the category that gives my friends on the far right and far left the willies: road funding. In the United States, road funding is primarily derived from a gas tax levied at the federal and state levels. Even if my modest prediction of EV deployment noted above comes true, there will be a significant decline in gas tax revenues, something that has, in fact, obtained over the past few years as internal combustion engine vehicles became more efficient and people generally driving less. EV owners will need to pay for roads, just like everyone else.

Okay, so what are we to do? There is really only one solution: charging users of roadways based upon vehicle miles driven and potentially other factors such as individual income levels and classification of vehicles. The close corollary is, of course, tolling. We toll roads, bridges, and tunnels in multiple jurisdictions all over the world. It works. For the future, however, we will need to evolve from ungainly gantries and ugly toll tags glued to windshields, especially as we need a ubiquitous funding mechanism.

Connectivity between the vehicle and infrastructure is the answer, and in 20 years, this will happen in multiple ways, including leveraging smartphones, systems embedded in vehicle electronics suites, as well as computer vision cameras. To be clear, your vehicle's movements will be tracked, and you will pay a road user charge based upon where you go. I hope that the price you pay will be calibrated to your income level and kind of vehicle you drive as noted above. This is an important improvement on the gas tax, which is inherently regressive in nature and effect.

In my mind, I can perceive ritual wailing and gnashing of teeth from all sides about privacy. My response is a question (okay two questions): do you own a cell phone and is it on? If so, get over yourself. Big brother knows where you are. In 20 years, we will be monitored by cameras for many reasons in many more places. I predict that by then a victim of a violent crime on a city street will be outraged if the perpetrator is not identified in a video.

Transparency and education will be the keys to public acceptance of road user charging. Policy makers will need to be explicit about the use of funds to build and maintain roads and perhaps to support other transportation modes such as transit.

Perhaps incongruously, I believe mobility-on-demand and mobility-as-a-service will remain elusive. The notion that one will be able to pay for "mobility" in some regular fee spanning multiple modes of travel available at an instant is difficult to imagine in this time frame. As those who have tried to organize this have learned, there are too many competing motivations among the public and private transportation providers, and it is hard to see these going away.

Again, all is not lost. I predict public transit will make great strides in 20 years with improved service at all levels, especially with respect to microtransit and addressing the needs of travelers for the first-mile/last-mile—or, more appropriately, the first-10-miles/last-10-miles. In achieving this, I believe transit agencies will first figure out how to serve the disabled, elderly, and children, collectively our most vulnerable travelers, extremely well. By doing so, agencies will make everyone's experience on transit better.

Investment in public transit is a necessary expense for a civilized society, in which many people are denied access to the social determinants of health (i.e., health care and healthy food) and to workforce training and employment because they lack a reliable means of personal transportation. Cars are expensive to own and maintain, such that large swathes of the population simply cannot afford them. I suggest public transit is a means to break the cycle of poverty and ill health for many. Add congestion associated with one person in one vehicle, and the case in my view is made.

I come to my last and favorite mobility prediction, and it is about parking. Those who think future mobility will involve the transformation of parking spaces, lots, and structures into vast green spaces or that parking requirements for new construction will be reduced or even eliminated will be very disappointed. First, if there is a higher volume deployment of fully automated vehicles than I predict, they will need a place to park lest empty robot cars roam the streets adding to congestion. And those EVs will need a place to plug in as widespread inductive charging capability will take a lot longer than 20 years.

Parking will be more automated for sure, allowing more vehicles to be accommodated in tighter spaces. Before this becomes ubiquitous, however, I predict some challenges in parking structures as the fully automated vehicles fight for spaces, but collision avoidance systems should prevent fender benders.

More good news: I predict people will be able to pay for mobility services much more easily than today. We may not reach "one account nirvana" in 20 years, but I believe customer accounts for transit, parking, and road user charging will be linked, perhaps using blockchain. Agencies that fear losing "stored value" in customer accounts to another agency will be mollified by the distributed ledger's transparent and dynamic qualities.

Customer service will need to be superlative with the latest financial technology applied to transaction management and payments processing. Back-office platforms will entail fusing transportation with the financial services industry to provide a superior customer experience for travelers. Be cheered by the fact that you will be able to identify, reserve, and pay for parking and charging your EV and for road use and your transit ride in transactions that will be almost frictionless.

In sum, the future of mobility in 20 years depends on realistic applications of technology coupled with firm commitment to change from the traveling public. Some things, like linking customer accounts, require no new technology—you could use an abacus and paper ledgers. Almost all aspects of future mobility will require circumspection and leadership from policymakers and participants to embrace and promote change. When will all of this happen? Time will tell.

About John Peracchio

John Peracchio provides strategic consulting services in the intelligent transportation systems ("ITS") sector and automotive industry for strategic and investor clients. His focus is on the deployment of mobility solutions across transportation modes, especially those involving connected and highly automated vehicles as well as innovation in transaction management and payments processing.

From February 2017 to April 2020, Mr. Peracchio served as chair of the Michigan Council on Future Mobility having been appointed by Michigan's governor. The council within the Michigan Department of Transportation provided policy recommendations to promote the development of technologies for autonomous, automated, and connected vehicles and to enhance personal mobility. Currently, Mr. Peracchio serves as a senior advisor to the succeeding Michigan Council on Future Mobility and Electrification and leads the council's electrification workgroup.

Mr. Peracchio has been active with the Intelligent Transportation Society of America and currently serves on its Standing Advisory Committee on Smart Infrastructure. He is a member of the Board of Directors of Feonix - Mobility Rising, a nonprofit organization dedicated to serving the transportation needs of underserved and underprivileged communities.

Mr. Peracchio received his Juris Doctor degree from Columbia University School of Law and a Bachelor of Arts degree from Brown University.

16

Aishwarya Raman

Director and Head of Research
Ola Mobility Institute

I spent my childhood in three cities across three regions in India. The cities differed from each other not only in their population size and degree of modernization but also in their language, culture, and outlook on life in general. What was a common fixture, though, of my urban life as a child in Bengaluru, Chandigarh, and Madurai was the humble three-wheeled scooter-rickshaw, commonly called auto-rickshaw, auto, or tuk-tuk in India and elsewhere.

My earliest recollection of stepping outside the safety and comfort of my home involved my family and me cruising in an auto-rickshaw through the garden city of Bengaluru in the early 1990s. Gulmohar, Amaltas, Jacaranda, and Bougainvillea trees lined our frequent auto-rickshaw journeys to the many parks in the city. The open rickshaw allowed the gentle breeze to caress us while the flowers often fell in our laps, blessing us with limitless joy only a child knows to experience.

It was in the auto-rickshaw that I made my earliest and closest friends. I took a tuk-tuk to school every day in each of the cities we lived in. The auto was not mine alone. I shared it with my schoolmates—those younger and older than me, and very few from my own grade. Each auto could accommodate anywhere from four to eight kids, depending on their grades and the size of the modified school auto. From kindergarten to grade twelve, my tuk-tuk friends became my extended family—we celebrated each other's birthdays, we wished each other the best for exams, and were sad to approach the end of the school year lest we forgot each other and made new friends during the summer break!

It was in the auto that I learnt the vernacular of each city I grew up in—Kannada, Hindi and Punjabi, and Tamil. I lacked the necessary opportunity to be fluent in the vernacular at school or at home—given how English was the medium of instruction in each of the schools I attended, and the language I speak at home is entirely different—Telugu. Thus, every time we shifted cities, my new set of auto friends, including the auto driver, taught me the local language and culture and made me feel welcome in this diverse, beautiful country I call home—India.

It came as no surprise to those close to me that for my undergraduate thesis, I studied the auto-rickshaw sector in Chennai. Notwithstanding my lifelong fascination with the three-wheeled automobile, I was intrigued by the informal nature of its operations, the informational asymmetry plaguing the auto-rickshaw industry, and archaic policies and regulations that stymied the growth of this sector. The sociological analysis earned me a coveted seat at the University of Oxford. Soon, armed with an MSc in Sociology, I converted my thesis on auto-rickshaws into a social enterprise, allowing me to impact the lives of 2,500 drivers and their families in South India in 2012-2014.

The first-of-its-kind enterprise was a call-an-auto service, which operated auto-rickshaws on meter with scientifically determined fares, effectively eliminating the lack of trust between passengers and drivers, and systematically improving the earnings of drivers, and the experience of passengers. I also created India's earliest all-women auto-rickshaw fleet and found creative ways to engage marginalized members of our society—transgender women. In December 2014, I—then the CEO and co-founder of an auto aggregator, a rare woman mobility entrepreneur, all but 24 years old—found myself as a lead discussant at a policy deliberation chaired by the honorable Transport Minister of Tamil Nadu. Along with others around the policy table—all experienced men of the industry, civil society, and the government—I helped Tamil Nadu, one of India's most industrialized, urbanized states with a large auto manufacturing base, update its auto-rickshaw fares and, without my realizing, helped usher in a new era of mobility and mobility governance in the country.

By my 25th birthday, I knew the mobility sector was my calling. Indeed, 2015 was the dawn of an unprecedented mobility revolution in India. Just 24 years earlier, India had embarked on a new economic journey focused on Liberalization, Privatization, and Globalization resulting in a rapid expansion of the country's economic output, catapulting India to the position of one of the world's fastest-growing major economies. This laid the perfect foundation for India to harness the power of the Fourth Industrial Revolution. As many as 243 million Indians became mobile phone internet users in 2015,[1] paving the way for app-based mobility aggregation to transform how we work, move, and live.

It was at this juncture that I joined one of the world's largest shared and electric mobility platforms, Ola. In its fifth year of operation, Ola was already a household name in major Tier 1 and Tier 2 markets. Keen to address the needs and aspirations of the next few million users and drivers, I helped Ola scale its auto-rickshaw service to nearly 80 cities in India within two years of joining the platform. The rapid digitalization of India and the entrepreneurial energy of young Indians made this possible. Today, there are over two and a half million auto-rickshaw, bike-taxi, and taxicab drivers affiliated to Ola who use smartphones to connect with Ola's hundreds of millions of customers present in over 250 cities across 4 countries.[2]

---

[1] "Number of mobile phone internet users in India from 2015 to 2018, with estimates until 2023." Statista. June 29, 2021. Accessed on August 10, 2021. https://www.statista.com/statistics/558610/number-of-mobile-internet-user-in-india/

[2] "Ola enables in-app 'tipping' globally, to help customers appreciate their drivers." Ola. June 2020. Retried on June 30, 2020. Accessed on August 10, 2021. https://www.olacabs.com/media/in/press/ola-enables-in-app-tipping-globally-to-help-customers-appreciate-their-drivers

This mobility transformation that India is witnessing powered by all things digital[3] is just the beginning. Just as I've been an auto-rickshaw user as a school-going child and a bus user as a college student, most Indians rely on shared mobility—public transit and intermediate public transport[4]—and walking and cycling for their everyday commute.[5] India also ranks low on private vehicle ownership with 23 cars per 1000 population, and 128 two-wheelers (mopeds, scooters, motorcycles/motorbikes) per 1000 population, compared to advanced economies witnessing over 400-800 cars per 1000 persons.[6] This makes India well-poised to ride the wave of shared mobility.

Today, the ownership of the ride (i.e., accessing a motorized or non-motorized asset) is replacing the ownership of the drive (i.e., owning a car).[7] Shared mobility, thus, limits the unfettered growth of motorized transport and incentivizes the switch towards public transit. With shared mobility platforms reaching India's hinterlands within a decade of their launch,[8] the day is not far away when every Indian uses a smartphone to plan and complete their everyday commute. This would mark the full adoption of digitalized shared mobility which is safe, accessible, reliable, and affordable for all.

Therefore, for me, the future of mobility is hinged on equity. It is about maximizing the utilization of an existing vehicular asset—thereby reducing congestion in a rapidly urbanizing India, making mobility affordable for all, and unlocking remunerative opportunities in the millions. The future of mobility is about democratizing access to economic opportunities—both for the users and those who are and aspire to be service providers in the burgeoning mobility economy. It is about making cities safer and readily accessible to persons with disabilities, women, the elderly, and children alike. It is about reducing the negative externalities of mobility such as air pollution and congestion and improving positive health outcomes of denizens. The future of mobility is about making cities livable.

The future of mobility is also about India staying ahead of the curve and becoming a global hotspot for mobility innovation. Today, new-age entrepreneurs are producing world-class cutting-edge mobility solutions in India for the world. India—having one of the world's largest and youngest populations—provides a fertile ground for accelerated testing and adoption of mobility innovations. Implementation in India will provide entrepreneurs a blueprint for successful implementation anywhere in the world. Thus, emerging business models and changing consumer needs and aspirations are bringing

---

[3] Today, there are approximately 470 million mobile phone internet users in India. https://www.statista.com/statistics/558610/number-of-mobile-internet-user-in-india/

[4] Intermediate public transport (IPT) refers to public transit of the smaller form factor taking the shape of two-wheeled taxis (bike-taxis), three-wheeled taxis (auto-rickshaws), and four-wheeled taxi-cabs, in addition to other shared mobility options such as vehicle rentals or vehicle pooling using two-wheelers and four-wheelers. IPT also includes micro-transit, i.e., mini-buses.

[5] "B-28 'Other Workers' By Distance from Residence to Place of Work and Mode of Travel to Place of Work—2011." Census 2011. Government of India. Accessed on August 10, 2021. https://censusindia.gov.in/2011census/B-series/B_28.html

[6] "Personal car per 1000 population." Knoema. Accessed on August 10, 2021. https://knoema.com/mitveqb/personal-car-per-1000-population

[7] "How the ride will replace the drive." Mint. March 15, 2019. Accessed on August 10, 2021. https://www.livemint.com/technology/tech-news/how-the-ride-will-replace-the-drive-1552585130633.html

[8] "Ola makes inroads into India's hinterlands with bike taxi service." Business Standard. November 2019. Accessed on August 10, 2021. https://www.business-standard.com/article/companies/ola-makes-inroads-into-india-s-hinterlands-with-bike-taxi-service-119112501468_1.html

about a paradigm shift in mobility. The future of mobility in India is shared, connected, electric, Artificial Intelligence (AI)-powered, and autonomous. And the future is here.

India, today, is not only a shared mobility leader but also an electric mobility powerhouse. Despite the COVID-19 pandemic, electric vehicles (EVs) have seen a rapid uptake these past two and a half years.[9, 10] Concerns around range anxiety, high upfront cost of EVs, etc. are being addressed by the public and private sectors jointly. Fiscal and non-fiscal incentives such as robust charging networks in cities and on the highways, purchase subsidies, low electricity tariffs for charging EVs, no permit or registration costs, low to no parking fees or toll, innovative and efficient electric vehicles, etc., coupled with the high cost of petrol and diesel making owning and using conventional vehicles prohibitively expensive, are accelerating the adoption of EVs in India.

India is also home to the world's largest electric two-wheeler manufacturing plant, Ola FutureFactory, located in the South Indian state of Tamil Nadu. The government and the industry alike are prioritizing the integration of renewable energy with mobility both on the fronts of manufacturing[11] and charging of EVs, as well as the proper end-of-life management of EVs and batteries. The Union government and select states are catalyzing the adoption of e-mobility further by offering incentives to vehicle owners to scrap their internal combustion engine (ICE) vehicles reaching their end of life and replacing them with EVs.

These are but a few of the strategic measures that the government and industry are taking[12] to ensure clean mobility is not merely the aspiration of the elites to own an electric car, but it is the reality that the masses experience every day. My vision of the future of mobility, shaped by India's realities and my own experiences, involves mass adoption of electric two-wheelers (after all, India is a two-wheeler country[13]) alongside the electrification of buses, intermediate public transport, hyperlocal delivery, and logistics fleets. This way the benefits of electrification—reduced air pollution, lowered cost of mobility, cities that are livable—accrue to large sections of the population at once and also reach the most vulnerable populace.

The future of mobility also involves embracing technological innovations that will make our lives better. Consider the curious case of AI in mobility. AI is not only about building driverless luxury cars but creating real-world impact.[14] In Kenya and Rwanda, AI-powered drones deliver vital medical supplies to far-flung hospitals. Delivery of food, medicines, and other essentials using drones and driverless (autonomous) vehicles is a recurring topic today

---

[9] "Opinion: Despite the pandemic, how can India accelerate the adoption of EVs?." ET Auto. September 2, 2020. Accessed on August 10, 2021. https://auto.economictimes.indiatimes.com/news/industry/opinion-despite-the-pandemic-how-can-india-accelerate-the-adoption-of-evs/77886245

[10] "More electric vehicles hit the road in 7 months than all of last year." The Economic Times. August 9, 2021. Accessed on August 10, 2021. https://economictimes.indiatimes.com/industry/renewables/more-electric-vehicles-hit-the-road-in-7-months-than-all-of-last-year/articleshow/85143510.cms

[11] As can be seen in Ola FutureFactory. https://olaelectric.com/futurefactory

[12] "EV-Ready India Part 1: Value Chain Analysis of State EV Policies." World Economic Forum in collaboration with Ola Mobility Institute. October 2019. Accessed on August 10, 2021. https://olawebcdn.com/ola-institute/EV_Ready_India.pdf

[13] "The Power of Two Wheels—Bike-Taxis: India's New Shared Mobility Frontier." Ola Mobility Institute. March 2020. Accessed on August 10, 2021. https://olawebcdn.com/ola-institute/bike-taxi-report.pdf

[14] "Harnessing AI in Mobility to Create Real-World Impact." Salzburg Global Seminar. November 2020. Accessed on August 10, 2021. https://www.salzburgglobal.org/news/latest-news/article/harnessing-ai-in-mobility-to-create-real-world-impact

amidst the COVID-19 pandemic in India and many parts of the world. The AI application of Advanced Driver Assistance Systems (ADAS) is fairly common in new-age cars in India as well as in taxis affiliated to platforms like Ola. AI is also deployed in public transit buses in the states of Uttar Pradesh and Karnataka to alert sleepy drivers and avoid collisions. Ola further uses AI to enhance personal safety and security of its customers and driver-partners by detecting route deviations and predicting and preventing unsafe incidents.

For India, the application of AI in shared mobility, particularly public transit, is game-changing. In a concept paper by the Ola Mobility Institute,[15] my colleagues and I build a case for revolutionizing India's mass transit system by harnessing the power of AI. We show how AI could be leveraged to make mass transit more demand responsive, financially viable for the private and public sectors building and operating it, and ultra-affordable for the masses.[16]

Such innovations are the need of the hour. As one of the most populated countries in the world coming of age with one of the lowest vehicle penetration rates of a major economy,[17] India has a unique opportunity to establish a sustainable mobility system that is safe, accessible, reliable, and affordable.[18] In 2047, when India would complete 100 years of independence, I see my compatriots experiencing travel delight, in urban and rural India alike. Universally accessible and affordable digital applications help them plan their journey, providing them with a range of mobility options to choose from in a dynamic, real-time fashion. AI-powered, electric, and demand-responsive public transit is well integrated with first- and last-mile motorized and non-motorized transport modes, both in the real world and digitally. Clean shared mobility powers most of the rides in a city, town, or village. Digital payments or, in other words, cashless travel is the norm. The only private vehicles are all-electric. Hyperlocal deliveries and long haul too are powered by zero-emission vehicles. The country harnesses the power of AI to make mobility systems including infrastructure—traffic lights, for instance—smart, and efficient. Such is the future I envisage.

The last two decades of the twenty-first century witnessed seismic shifts in mobility, catalyzed by the ubiquity of smartphones, availability of data, and most importantly concerted efforts towards minimizing negative externalities of mobility: traffic congestion, air pollution, and energy inefficiency. The human and economic costs stemming from these externalities must be overcome with incisive, affirmative, ambitious, and strategic action. India is showing the way to leverage the new paradigms of mobility—Shared, Connected, Electric, AI powered, and Autonomous, thereby reinventing how mobility is conceptualized and delivered, effectively addressing many mobility challenges globally. Indeed, mobility is the new economic revolution, and India is the new frontier.

---

[15] Ola Mobility Institute (OMI) is a new-age policy research and social innovation think tank, focused on developing knowledge frameworks at the intersection of mobility innovation and public good. The Institute concerns itself with public research on electric mobility, energy and mobility, urban mobility, accessibility and inclusion, and future of work and platform economy. All research conducted at OMI is funded by ANI Technologies Pvt. Ltd. (the parent company of brand Ola). https://ola.institute/

[16] "ATOM: AuTonomous On-demand Mobility—A first-of-its-kind AI-led public transport concept for India and the world." Ola Mobility Institute. October, 2021. https://olawebcdn.com/ola-institute/atom-report.pdf

[17] "Beyond Nagpur: The promise of Electric Mobility—Lessons from India's first multi-modal e-mobility project." Ola Mobility Institute. April 2019. Accessed on August 10, 2021. https://olawebcdn.com/ola-institute/nagpur-report.pdf

[18] "Gaps in mobility network." The Business Line. May 24, 2021. Accessed on August 10, 2021. https://www.thehindubusinessline.com/opinion/gaps-in-mobility-network/article34635982.ece

About Aishwarya Raman

Aishwarya Raman is Director and Head of Research at the Ola Mobility Institute (OMI), a new-age policy research and social innovation think tank developing knowledge frameworks at the intersection of mobility innovation and the public good. An MSc in Sociology from the University of Oxford, Aishwarya has over a decade of professional experience in the mobility domain. She is the co-founder and ex-CEO of AutoRaja, one of India's earliest and largest book-an-auto services. She also created India's earliest all-women auto-rickshaw fleet, AutoRani. She started her journey with Ola as the head of the Ola Auto category across North India in 2015. Today at OMI, Aishwarya conducts research in areas such as sustainable urban mobility, electric mobility and energy, gender, accessibility and inclusion, future of work and the platform economy, AI, and much more. Aishwarya is a member of the Global Future Council on Urban Mobility Transitions at the World Economic Forum. She is a Salzburg Global Fellow participating in the Japan-India Transformative Technology Network. Aishwarya advises and mentors organizations, researchers, and young professionals, including the Global Partnership for Informal Transportation and Young Leaders for Active Citizenship, among others. In her previous avatar, Aishwarya was an academic, teaching Sociology at the University of Madras, India.

Courtesy of Aishwarya Raman

Courtesy of Karina Ricks

Karina Ricks

Director of Mobility and Infrastructure
City of Pittsburgh, Pennsylvania

Transportation is personal. I hear it all the time when I talk to people all across the country. I hear about it less from higher-income households. Sure, these workers and their families may complain about congestion or fantasize about robot cars or flying taxis, but as a matter of daily life, how they will travel doesn't take much thought. They have the ability to jump into their car or hail or borrow a shared ride without scarcely a thought. I know because I am one of them.

For others—those with disabilities or lower incomes—transportation is often at the forefront of their mind. They need to plan out travel carefully, and every trip is a game of Russian Roulette. If a low-wage worker is able to scrape together enough for a monthly bus pass, she likely doesn't have any funds left to take a scooter or cab if she misses her bus. A low-income, rural household may have an automobile, but they are one flat tire, one car repair, or one parking boot away from total isolation.

I have heard so many stories like this over my career. Every day they continue to challenge and inspire me to think, "What are the real problems that need solving?" and how do we make our mobility system better.

Several years ago, in a Midwestern city, I met a man at an outlying bus stop. We were intercepting regular people around the city to ask, "What is your big idea for transportation?" In busy downtown lunch spots or university centers, we heard ideas for hyperloop, subways, and smart parking. But as this man rode up on his bicycle in his landscaper uniform, he said, "More bike racks on buses." Initially, that didn't seem like a really big idea, but he explained. "If the bike rack on the bus is full, I either have to wait here for another bus with space in the rack, or leave my bike and transfer downtown to get to work, but either way I am twenty minutes late.

My boss likes me, but says he won't promote me when I'm late, so bigger bike racks would make a big difference to me."

Another time in another city, I met a young woman at job placement center and asked her the same question. "Any-direction bus transfers," she said. "I have to take an outbound bus to drop my kids at daycare, but all the good jobs are downtown. I can't afford to pay for two extra transit rides a day, so even though I could get a good downtown job, I can't afford to take it. I can only take a job I can get to with the free transfer, and there aren't many, so I am still unemployed."

While working in Washington, DC, I met a single mom from a southeast neighborhood. Although she lived only a few miles away from her office job, congestion made her commute trip unpredictable. Most days the bus trip would take only 30 minutes, but often enough it was ninety minutes or more. She worried she would lose her job if she was late and didn't want to have to start all over somewhere else. So she leaves an extra hour early every day and has to leave her school-aged kids home alone for that hour to get themselves out of the house and to school on their own. She told me how she worries every day for their safety, but doesn't really have an alternative.

I met another mother who described how every day she gets up at 5 am and gets her children out the door by 6:00 so they can start their daily routine. She has to take five buses to get her four kids to three different schools and herself to work—two and a half hours of travel twice a day! The family's travel day doesn't end until after 8 pm most days, leaving her no time but to grab fast-food dinner and forcing the kids to do homework on the bus or waiting for a transfer on their lengthy ride home.

The stories go on and on. Real, lived experiences that illustrate the mobility insecurity so many aspiring Americans struggle with day after day—a restaurant worker who spends a good chunk of her tips every night for a ride home lacking other late-night options, a first-generation college grad who turned down his dream job because transit just doesn't go there, a returning citizen who can't get work because he can't get a driver's license, and a mom who shops at the local gas station because she cannot get groceries any other way.

All these stories are a common story of system failure—or, rather, a failure to build a system. Transportation is the keystone in our economic, health, and education systems. Failures in the transportation system undermine and destabilize all other dependent systems.

The good news is we can address this. We know how to build a system of systems. We have the tools and know-how to integrate mobility services. We know this will provide greater resiliency and efficiency across systems and greater reliability for travelers. Furthermore, we can take away mobility insecurity by providing universal basic mobility for all. Not only will this provide a firm foundation for individuals to pursue economic mobility, but it is the firm foundation for a growing national economy.

Whether two years or two generations in the future, the future of transportation must be a future of integrated systems. Micromobility, hyperloops, autonomous vehicles, and unmanned aerial travel all hold promise in filling mobility gaps, reducing transportation emissions, improving convenience, and increasing safety. However, if each emerging new Mobility-as-a-Service remains a proprietary, siloed system, we will fail to optimize the opportunity of holistic mobility.

The future of mobility is one where transit is understood to be more than buses and trains, but a system of mass mobility that incorporates bikes and scooters, autonomous shuttles, and whatever comes next. It is a system that not only brings people to destinations but also destinations to people, incorporating deliveries and data communications.

It is not a parochial public pursuit, but a partnership among public, private, and nonprofit providers. It is a well-governed system that provides the accountability, transparency, and privacy protections essential to public trust.

The future of mobility is one where "customer experience" is not something hawkishly protected by individual providers, but a common platform legible to those with even the lowest digital literacy or access. It is a system where the traveler only needs to figure out where and how she needs to travel, and the system itself can sort out on the back end how that single fare for a multimodal trip is allocated among providers.

The future of mobility is one where we invest in and, yes, subsidize the system and services in proportion to the value they return to our society and economy. That value is great and social impact investing means we look beyond a single column or sector in the economic ledger. That gentleman with his bike? Think of the lifetime of foregone earnings that he is denied where, through his economic multiplier effect, we would more than recoup an investment in intermodal integration. The woman at the job center would happily stop drawing unemployment benefits if we invest in transit transfers instead. We can pay it forward a generation or two with smart signals that provide greater reliability in transit travel time and give parents more time at home. Guaranteeing basic mobility for the mother of four can simplify her complex commute and improve nutrition, learning, and health for her family.

I am optimistic about the future of mobility, if we move beyond a mere fascination with whiz-bang technology, shiny vehicles, and flashy apps and build that system of systems. We need to make investments that blunt the cutthroat competition among mobility providers who have yet to see a profit and compensate these services for the benefits they provide instead. We need to establish workable mobility governance systems anchored in principles and priorities for the public good.

We can do this. The future of equitable transportation and economic mobility can be days, not decades, away if we listen to today's lived experiences and remain laser-focused on the real problems that need solving.

About Karina Ricks

Karina Ricks currently serves as Director of Mobility and Infrastructure for the City of Pittsburgh, Pennsylvania. In this role she both works with new technology innovators in AI and robotics and manages old technology infrastructure of legacy bridges, streets, and public steps. Prior to moving to Pittsburgh, Karina worked in and for the District of Columbia—first as Director of Transportation Planning and later establishing the mid-Atlantic office of Nelson/Nygaard Associates. Karina has worked at the federal and international levels as well as on issues of environmental protection, economic development, and democracy building. The daughter of immigrants and a mother of two, Karina is strongly motivated to improve equity and inclusion and ensure the future of a sustainable planet. A lover of the outdoors, Karina is often found out on the trails of urban parks or satisfying her curiosity about places around the globe. She is most proud of the individuals she has mentored and cultivated over her career, many of whom are leading change across the nation today.

Alex Roy

Founder
Autonocast

D o you like driving? I do. That's why I founded the Human Driving Association, which started as a stunt but grew because it unexpectedly tapped into something much deeper, which is the fear that self-driving technology might someday replace human driving. But the more one knows about history, the more one realizes this is nonsense.

Driving will thrive, not despite technology, but because of it.

To understand why, we have to look at the true history of technology, after which the real questions emerge: What do "driving" and "self-driving" mean? What do we actually want? And what technologies might we yet build to get there?

The Self-Driving Litmus Test

SAE's automation levels have long been misinterpreted. To the end user, there are only two types of vehicles: human driven and self-driving. The litmus test? A UX test I call Roy's Razor: can you pick a destination and sleep in the back? That's self-driving. Everything else is human driven, no matter how much Advanced Driver Assistance (ADAS) is present.

The True History of Technology

Every time new technology is developed, two narratives emerge. The tech-utopians claim X will replace Y and the past deserves to die. The techno-pessimists claim X can never work, and even if it did, no one will or should want it because the old ways are better.

This happened with the printing press, electricity, trains, and planes, and it's happening now with self-driving technology.

Neither narrative makes sense. Every debate around progress oscillates wildly around a single truth climbing inexorably up the arc of history: progress is inevitable. We can't stop it, but we *can* shape it, and nothing requires us to choose between two extreme futures defined for us.

Self-driving is inevitable, but what is not inevitable is that it conflicts with human driving. Automation is not zero-sum. Binary thinking is slavery to a lack of imagination. Elevators didn't replace stairs, they supplemented them. Elevators automated the climbing we didn't want to do and enabled taller buildings, denser cities, and the enormous economic and cultural growth that came with them.

But there was still more innovation to come because, if you wanted to move a lot of people a short distance, an elevator wasn't necessarily the best solution, a hybrid was. Enter the "Motorstair" or "Electric Stairway," which today we call escalators.

One hundred and twenty years into the era of automated vertical mobility, elevators and escalators still haven't replaced stairs because the freedom of choosing to use our own two feet isn't a problem to solve, it's a fundamental human right no one would ever give up.

What Do Driving and Self-Driving Actually Mean?

Driving has come to mean two things: commuting, which no one enjoys, and *driving*, which many do. Self-driving is one potential solution to commuting, but it can never "solve" driving for the same reason we still choose to walk and run and climb mountains.

But self-driving can solve more than just the work of commuting; it has the potential to make our streets safer anywhere they operate, not just for passengers but for cyclists and pedestrians too. Self-driving vehicles won't get drunk or tired, and their sensors can already see farther than we can, in darkness and bad weather, in 360 degrees.

What Do We Actually Want?

Everyone wants the freedom to drive anywhere, anytime, *and* the safety of self-driving, of course. The freedom part is easy, but self-driving will never extend to 100% of the places and conditions in which humans drive. Why? Because self-driving technology will—by design—never take the kind of risks humans do, which means there will always be places and times *only* humans will drive.

So how do we improve human driving safety while preserving the freedom to drive?

We reconceptualize ADAS as we know it. We need more than better cars, we need better drivers, and the best way to train them is in the cars they drive.

How Driver Augmentation Will Transform Driving

As of 2021, every new car sold in the United States has some form of ADAS, but even the best systems have three fundamental limitations. First, ADAS is only designed to mitigate the consequences of human error, whereas self-driving is designed to prevent it. Second, ADAS doesn't explain itself; few understand what it can (or can't) do until it's too late. Third, ADAS doesn't improve driving skill, which is why suboptimal drivers often make the same mistakes over and over, even after they get into a crash.

By transforming driver assistance into driver augmentation, we can create a new generation of driving enthusiasts, and make driving safer than it's ever been.

Luckily, all the enabling technologies required to evolve ADAS into a driver augmentation system already exist, at least in prototype form. Sensors, data sharing, connectivity, and maps are table stakes, so I've excluded them. Let's look at how a fresh set of future technologies will converge to transform driving:

Electrification—EV powertrains are already commoditizing performance previously available only in gasoline-powered hypercars. Safely driving them requires skills taught only in professional racing schools, which is not feasible or cost-effective.

Historic Vehicle Performance Databases—Legacy brands will become *more* valuable, as new car buyers will want to experience vintage car performance, sound, and handling via downloadable content (DLC), just like sim racers have done in-game for 10+ years. Simply download different car profiles to your EV, which will tailor the experience to the car of your choice.

Gamification/Education—Driver education will become in-car edutainment. This was solved 10+ years ago in driving sims like Forza Motorsport, Gran Turismo, and iRacing, which unlock cars/performance as players pass in-game driving tests.

Augmented Reality + Head-Up Displays—Both technologies are approaching sufficient maturity for projection of HD real-time data onto displays, which will eventually encompass the entire windshield, greatly improving driver situational awareness.

Camera-Based Driver Monitoring Systems will become **Cognition Management Systems**, personalizing vehicular performance to individual driver situational awareness and skill in real time.

Dynamic Driving Envelope Protections—inspired by aviation flight envelope protections, the boundaries of vehicular performance will dynamically expand or contract based not only on cognition and conditions, but location and proximity of other cars, bikes, and pedestrians.

It is impossible to predict when the sum of these enabling technologies will deliver the dream, but we can already see the contours of their convergence in statements from companies like Porsche and Rimac, who have hinted at future in-car track driver education platforms, and Toyota, whose Guardian project seems like a precursor to what I hope driving will become.

And let's not forget the final prize: dual-mode vehicles. Someday self-driving technology will be available in the same amazing cars we drive, but that's a ways off. Until then I'd rather summon a robotaxi to go downtown. I don't want to deal with the parking.

About Alex Roy

Alex Roy is Director of Special Operations at Argo AI, host of the Autonocast & No Parking podcasts, editor-at-large at the Drive, founder of the Human Driving Association, and Producer of APEX: The Secret Race Across America. He held the Cannonball Run record from 2006 to 2013. The personal views expressed herein are solely his own and do not reflect those of Argo AI.

19

Courtesy of Avinash Ruguboor

Avinash Rugoobur

President
Arrival

Transportation is a key factor for personal well-being and economic development globally. So it has always blown my mind that as you travel the world, you see the same vehicles everywhere regardless of how the town, city, or village operates. Wealthy nations are losing productivity and increasing stress as people spend many hours every day stuck in traffic using a transportation system with ever-escalating running costs and unacceptably high numbers of lives lost in accidents. As you travel to areas of less wealth, we see these problems exacerbated by vehicles that are "hand-me-downs" of designs made for wealthier nations—polluting buses, old trains, and cars that aren't fit to be on the road. Additionally, according to the EPA, transportation is the largest contributor of US greenhouse gas (GHG) emissions at about 29%(!) of total GHG emissions. When you take the time wasted, lives lost, and enormous health impacts from the pollution of the current system and add the risks of climate change, which does not discriminate on who or where you are, it is clear we must do something about this now. It doesn't matter how we got here, what matters is learning from it, making a change, and putting it into action now. Our next move is critical for our future, and thankfully, as I look around, I finally see mobilization and action and I am hopeful for that future.

I envisage the future of mobility to be not just about the technology we deploy but the principles that we use to design mobility solutions—clean, circular, and equitable. These interconnected principles lead us to solutions that are net-zero, recyclable, and reusable and which cater to all people everywhere, for all needs and tastes, regardless of socioeconomic background. We are no longer limited to a one-size-fits-all mobility system, but each city and community has the products that serve their needs, customized to how people and goods move around efficiently. For example, ambulances aren't converted vans, and we aren't using standard passenger cars for ride-sharing, we are producing green purpose-built vehicles designed for the very purpose they actually serve, making them efficient and affordable.

Clean applies to the vehicles themselves, but also the production system. We have been purposeful and thoughtful about how we source, build, maintain, and recycle vehicles to ensure zero net emissions and footprint. The production process is using renewable off-grid energy, reusing water, and is net-zero. The vehicles are obviously zero emissions, but we are powering them off renewable energy sources. Clean also extends to the process used to extract the materials and powertrains of the vehicles themselves. We are transparent in our supply chain to give the public an honest and open explanation of how their vehicles are made and what happens to them after.

Circular means that the vehicle, equipment, tools, and processes are reusable and recyclable. Modular architectures in both hardware and software enable the vehicle to not only stay in use longer but also keeps them up to date and provides an amazing experience for users throughout their usable life. The vehicles of the future are computer platforms on wheels, and when vehicles are finally at end of life, the components within them are used in different industries and don't go to landfills. The rest of the vehicle is recycled and feeds the system again.

Mobility is seen as a right for all, and equitable transportation is the norm. Public transportation is the foundational core of transportation around the world and sits alongside private and distributed ownership as an ecosystem. Users get quick and efficient access to public transportation with the network leaving no gaps and on-demand services and products enabling people of all backgrounds and abilities to travel anywhere, giving greater access to services and jobs. We no longer have two standards of transportation for wealthy nations and others. From city to city, we see transportation that has been customized for the use case, and roads are moving freely. Technology is used to optimize infrastructure with dynamic lanes, routes, and easy access to shared mobility to reduce congestion. Mobility-as-a-service is pervasive with the ability to access different forms of transportation across all ranges of the socioeconomic spectrum and use cases. This isn't for private citizens either, but the commercial sector is also using a vehicle pool that is optimized for the type of cargo and provides uptime across even multiple organizations. There's no wasted space and no wasted time.

These technologies extend to the infrastructure, which communicates with vehicles that ensures safety and increases efficiency by reducing the time spent in traffic. Idle parking lots are now optimized for dual usage, such as charging hubs or service and maintenance. We open up the ability to use the gig economy to service and maintain vehicles guided by technology such as augmented reality. Mobility in communities is not just limited to the roads but has expanded to the air, seamlessly connecting to the road infrastructure through hubs. Land, air, and sea vehicles are all powered by renewable energy sources.

To me, these principles set how the design and engineering teams of OEMs go about their work, how governments work together for urban planning, and how funding and incentives from both public and private sectors reward the development of these technologies. With businesses and governments working side by side to design products that meet specific needs and producing them in local markets to drive huge economic benefits in every region through job creation and utilization of local supply chains, we can further uplift the communities those vehicles are serving. These communities continue to progress as the economy transforms with transportation opening up opportunities to people who would otherwise be left out. We have made the full transition to moving people and goods through a clean, circular, and equitable mobility ecosystem.

To get to the future we want, we must start designing and committing resources to development now. This may be too optimistic, but it's worth working for. The world is changing, and it's time for mobility to catch up.

About Avinash Rugoobur

Avinash is President of Arrival, joining the company following a successful career at General Motors (GM), Cruise Automation, and his own award-winning ventures Curve Tomorrow and Bliss Chocolates.

As Arrival's President and Chief Strategy Officer, Avinash is responsible for Arrival's business and product strategy and international expansion. In his role, he oversaw Arrival's public listing in March 2021, which was the UK's biggest-ever IPO with a $13 billion float on Nasdaq. He also led the close of over $300 million in investment from blue-chip investors including Hyundai, Kia, UPS, and funds managed by BlackRock.

Avinash has international experience in leading and growing businesses across multiple segments. During his time at GM, Avinash was responsible for the ~$1Bn acquisition of Cruise Autonomous and part of the team that saw its subsequent valuation increase to $14Bn. This work was pivotal in accelerating the delivery of AVs as well as creating the OEM-Startup ecosystem.

Avinash has a passion for projects with positive social impact and is focused on bringing affordable zero-emission transportation to everyone.

20

Dr. Anthony M. Townsend

Urbanist in Residence
Cornell Tech

B ack in 2001, Steve Jobs was among the first to ride a new electric-powered scooter, the Segway. "If enough people see the machine you won't have to convince them to architect cities around it. It'll just happen," he told a reporter for *Time*.

Jobs was dead wrong. We all took one look at the Segway and walked away laughing. But the machine rolled on, making do with a niche filled mainly by mall cops.

Twenty years passed. Segways are still a footnote in the history of mobility. But we now know that Jobs saw the future clearly. Mobiilty is quickly becoming personal and electric. But even these developments didn't play out in precisely the way that Jobs and Segway inventor Dean Kamen imagined.

The first difference is that the pedestrian technology of batteries mattered more than Segway's brilliant gyroscopic stabilization. It turns out that people don't mind doing the work of balancing a scooter. But what they do mind is getting stranded with a drained power pack. And that's where a technology borrowed from mobile computing and lithium-ion batteries has made all the difference.

The second difference is the sheer variety of forms that personal electric vehicles are taking. Electric-powered bikes, scooters, and skateboards of all shapes and sizes now motor along your neighborhood streets. Still, this is not just a young person's game. Electric wheelchairs outnumber them all.

The biggest wildcard has been automation. The Segway's stabilization was a kind of automation. What's more, it showed that automation in personal EVs wasn't neces-sarily about full self-driving, but simply taking over difficult or costly tasks. The growth of micromobility has shown how much people enjoy cycling and scooting around cities. But rebalancing, or repositioning rides to match supply and demand, is a costly task for operators. Already, startups are piloting scooters that will simply drive themselves to the next customer. They're finding that because personal EVs are lighter and slower, there's less risk and complexity involved.

I have a name for this whole class of personal, automated, electric vehicles—*rovers*. And what fascinates me about rovers is exactly what Jobs speculated about when he saw the Segway. How will these things let us live? And how will we reorganize our communities around them?

The possibilities are easy to imagine, and many are already playing out around us. People that buy e-bikes often do so instead of a car, or a second car. They quickly find, too, that they can live further away from transit stops and downtowns. That helps them in the hunt for affordable housing and livable neighborhoods. But it turns out that delivery services are taking advantage of these innovations too. When everything can come to you on green wheels, there's less reason to go out to local shops and dining.

How will it all add up? My hunch is this. Personal electric mobility, automated or not, is going to give us new options for getting out of our cars and living closer together. But it will challenge all of our current assumptions and fashions about the best way to do that. Transit-oriented development and downtown revitalization schemes are all based on what we know about walkability (which is, sadly, quite little). Much of that is out the window when a large share of the population is zipping straight past main street to new housing tracts a mile or more out from the center. Just as we've figured out an antidote to exurban sprawl, a new kind of "microsprawl" could do our well-laid plans in.

And that's the point really. Jobs' laissez-faire approach to the future city isn't really acceptable. Good city design won't "just happen" because a new mobility technology has upended our lives. We need to be deliberate about anticipating the change and preparing for it.

About Anthony Townsend

Dr. Anthony Townsend is Urbanist in Residence at Jacobs Technion-Cornell Institute. As part of the founding team of the university's Urban Tech Hub, located on Roosevelt Island in New York City, he directs applied research and teaches courses on smart cities engineering. Anthony is the author of two books on the future of cities and technology, *Ghost Road: Beyond the Driverless Car* (2020) and *Smart Cities: Big Data, Civic Hackers and the Quest for a New Utopia* (2013), both published by W.W. Norton and Co. His consultancy, Star City Group, works around the world with industry, government, and philanthropy on urban tech foresight, policy, and planning studies.

21

Courtesy of Marla Westervelt

Marla Westervelt

Director of Policy
Commission on the Future of Mobility

t was a Friday afternoon in the summer of 2015 in Washington, DC. Like any summer afternoon at the end of the week, I was looking forward to leaving work and enjoying my weekend outside. Around 5 pm that day, my boss, Joshua Schank, walked into my office and said he would like to speak to me before I left. The visions of my Saturday morning bike ride quickly left my mind, traded in for anxiety regarding what was urgent enough to discuss on a Friday evening.

When I entered his office, he asked me to close the door. My anxiety was increasing. Was I about to be laid off?

Suddenly, a grin spread across his face. "I've been offered a leadership job at LA Metro to be the Chief Innovation Officer." I stared back. I wasn't being laid off. But my world was certainly about to change.

I had come to the Eno Center for Transportation about two and half years earlier. Eno is the leading U.S.-based national transportation policy think tank. I joined the team right out of grad school as a 10-week summer fellow and became so enamored with the work that I never looked back. At Eno, we were positioned to identify the big thorny issues in U.S. transportation policy, study them, and develop recommendations of how to overcome big challenges. For a curious young policy researcher, there was no better role.

Joshua continued, explaining that Phil Washington, who was the former CEO at the regional public transit agency in Denver, had moved to LA to be the CEO at LA Metro. We both knew Phil because he was on our Board of Directors and deeply respected his transformational work in Denver. One of the first things that Phil did upon joining the LA team was to create a new department, reporting directly to him, called the Office of Extraordinary Innovation.

As I was processing the information he said, "I'd like you to come with me, and lead our research team." I paused and thought. Me? In Los Angeles? As a Midwestern kid I had dreamed of moving to the East Coast, but never to California. And Los Angeles? It wasn't even on my radar of cities I thought I'd move to. But, nonetheless, without hesitation, I said "I'm in."

About eight months later, I got off the plane at LAX. I vividly remember how sunny that day was, not realizing that it was every day in LA. I headed to my new apartment and got down to business.

The first few months at Metro were a whirlwind. Coming from the world of high-level policy in DC to the world of implementation at a large transit agency was a culture shock. While in DC we spent a lot of time talking about how to fund programs or define optimal governance structures, in LA we were focusing on ensuring continued operational performance. When you're running an operational business, the ideals of how things should work or ought to work are high-class problems.

With naivete as the key tool in my arsenal, I set out to develop my workstream as the lead researcher. Without understanding the problems of the agency or the context, I dove in head first. I decided I wanted to understand how to build a productive partnership between a public agency and the private operator.

By this time, it was 2016. The ride-hail giants were reaching their peak—they had funding, they had a service model that put the customer first, but they didn't have a sustainable business model, and as we would all later find out, they were exacerbating congestion and air quality issues. I wanted to figure out how to capture the best that the public and private sectors had to bear.

Concurrently, the U.S. Federal Transit Administration (FTA) was scheming up a plan. They were developing a program to create flexibility to test out new programs and to provide seed funding called the Mobility on Demand (MOD) Sandbox program. This was my opportunity.

The grant was managed out of FTA's research department, which allowed it unique flexibility. For this type of grant, applicants were required to select their partners and vendors prior to submission. Applicants that were ultimately selected to be grantees would then be able to sole source all of the partners within their contract rather than be required to go through a lengthy competitive procurement.

While this flexibility was a great tool to get pilots on the ground faster, it meant that there was a lot more up-front work required to effectively develop a grant application. And with about two months between the release of the Notice of Funding Opportunity and the application deadline, there was no time to waste.

I worked with my team to get alignment on our concept. We envisioned identifying service gaps within the network that weren't easily filled by existing public transit service. Through the grant, we would leverage a partnership with a ride-hail company to deliver affordable, equitable, and accessible service in these gaps, supporting operation with a sustainable business model.

I picked up the phone and started making calls. To my delight, I was able to pull together a far-reaching team that included multiple transit operators in LA County, and researchers. What's more, we were also able to bring in two operators from King County, Washington to deploy an analogous project that would allow us to compare and contrast the pilot deployments. Finally, Lyft agreed to join the application as our operator. We hammered out the high-level parameters of our engagement, developed an application, and hoped for the best.

I had no idea what to expect—I had never developed a proposal for FTA and had no idea if my approach would be effective. About four months after our submission, in October, I got the call. We had been selected and granted $1.35 million, the largest grant of the cycle, to test out our concept.

By design, the project was going to be challenging to execute. There were a lot of partners (probably too many), there were two deployment regions, and based on time constraints in the application phase, there were a lot of details to work out.

The biggest of those messy details were (1) what data was Lyft going to share with the public agencies to help track the program and measure success, (2) how were rides going to be delivered for people in wheelchairs when there were very few wheelchair-accessible vehicles owned and operated by Lyft drivers, and (3) how the public agencies were going to define their business relationship with Lyft.

During the application phase, the team members had had discussions about each of the components, but as is typical, it became much messier when it was time to actually implement.

At the time, as is true today, the ride-hail giants had not yet turned a profit. For a passenger transportation service, this is not uncommon. Historically, it's really, really hard for passenger transportation operations to be profitable. And even harder (if not impossible) to make a profit if there is an expectation that the service will be affordable, equitable, and accessible. The best way to overcome these pitfalls is to partner with the government to subsidize operations.

Armed with this knowledge, my vision was to set up a framework to enable Lyft to earn a margin while ensuring they were meeting the needs of the public such as providing equivalent service to folks in wheelchairs and charging an affordable price. I envisioned that this project would help set the tenor for how public and private operators could work together and set the stage for a new business model for direct-to-consumer private operators.

But my vision was not shared by all. Our partners at Lyft had a different job to do. They were tasked with building scalable partnerships with governments and had already developed a specific methodology to execute such partnerships. My vision would have required a much greater investment than they were ready to make, in addition to the willingness to accept that their pre-existing business model was unlikely to be profitable on its own.

Ultimately, we couldn't drive alignment on our three key sticking points, and I was left to find an alternative operator. Via Transportation stepped up to the plate. With a business model focused on business-to-government software licensing rather than on direct-to-consumer operation, they were willing and able to work within the parameters we defined.

A key part of MOD's was the public and private sectors co-designing the service together from the beginning. In this case, it proved to be significantly more chal-lenging than I anticipated—but I believe as we continue to develop better contractual and regulatory frameworks, partnerships will become more productized. Traditional transit models in the USA, where the government distills a wish list into a request for proposals, commonly referred to as a Request for Proposal (RFP), are often ineffective as a tool to meet all of a transit agency's goals. This exercise necessitates that the private operators essentially already have a relationship with the agency and partners in order to be competitive. These prerequisites, and the inherent constraints within typical RFPS, often drive down the space for essential creativity and innovation. The future of mobility will need a framework where the private sector is effectively positioned to ask for what they need from the government to be able to effectively provide a service, and also critically be able to provide that service with a clear profit margin.

Through each of the roles I've had since leaving LA Metro, I've continued to be fascinated with the economic tension between public transit and private operation. LA Metro's project with Via was a success and it demonstrated what was possible. But it was just a start. There is still a significant amount of work to be done to assemble the pieces for a sustainable and replicable business model. In the future, I hope and anticipate we will develop even stronger software to plan and identify key service gaps and how to best fill them and predict changing demand and optimize routes. In this framework, governmental entities will need to continue to provide policy guidance, positioning themselves as the regional North Star.

I often look back on that DC summer day, and my audacity to jump in and take on the role at LA Metro, which fundamentally changed the way I see the industry and how I prioritize my time towards positive change. After doing a tour of duty on the implementation side, serving in both public and private sector roles, I'm back in a gig that lets me think big thoughts. After having worked at LA Metro, Bird, and at a non-profit that manages the data specifications that power Google and Apple maps, I'm more sure than ever that figuring out how to create sustainable direct-to-consumer operational models is a critical component to our future mobility success.

When I think of the future of mobility, I think of a robust, integrated mobility menu that makes it easy to select a mode other than your personal vehicle. Not only do I believe that this level of optionality would improve quality of life, but it is critical in order to enable us to meet our climate goals.

About Marla Westervelt

Joining the Commission as the Director of Policy, Marla has spent her career contemplating the intersection of the public and private sectors in driving towards mobility outcomes that are in the public interest. With a decade worth of experience in the transportation sector, Marla spent her early career working at the Eno Center for Transportation, where she led research efforts from reforming air traffic control governance to rethinking the way we pay for infrastructure in the USA. From there she headed to LA Metro, where she was a founding member of the Office of Extraordinary Innovation, leading research efforts. She then left to be an early member of Bird Rides, leading global data sharing policy and government-facing business development efforts. Most recently, Marla helped in the initial business development of MobilityData. This broad wealth of experiences uniquely positions Marla to drive a thoughtful and rigorous research agenda, exploring and seeking to define the future of mobility.

22

Candice Xie

CEO
Veo

How Micromobility Drives the Future of Transportation

'm not sure if every CEO has a precise moment when they knew they'd be dedicating their life to something bigger than themselves, but I did. That moment was checking out a bike from a clunky bike-share stall and being able to commute across town for my first job in Chicago. I was not bound by traffic or parking. The process was far from perfect but liberating, nonetheless, and I saw how it could be innovated upon endlessly. I hate owning a car, and I had been fighting it through my college years, but with my first job five miles away, I had finally caved. Then I discovered bike-share and was both excited and starstruck by the same vision many others had seen. A car-less cityscape. Streets lined with safe and clean micromobility lanes beaming with bikes, ebikes, scooters, covered pods, and endless other new and different micromobility devices to fit each individual personality, physical preference, or trip purpose, while keeping our air clean and our bodies healthy. I saw car lanes converted into outside seating for restaurants and community-building activities. Trees and cobblestone instead of parking spaces and asphalt canyons. And finally, seamless connection to public transit and other autonomous (and shared) travel options for longer trips. This vision probably doesn't sound new to you, but in most of the industrialized world, it remains far from the reality we experience on the ground.

There are huge challenges and costs for cities to transform our cityscapes and unleash micromobility demand. Infrastructure requires investment, and they need reliable partners to take the risk. Cities plan 10-20 years into the future, and my industry has a track record of companies that combust much faster than that. Driven by unfathomable amounts of cheap money, fueled with questionable business plans, unrealistic assumptions, and hockey stick business projections, they boomed and busted before my eyes. Many in this industry came from Silicon Valley and insisted that they were technology companies, not transportation companies, and promised software-like revenue growth. The damage done by those early players broke trust with cities and the public, making it harder than ever to move forward. The truth is we are in the transportation business, and that has always been defined by thin margins, heavy competition, and no room for waste.

So, while we all love daydreaming of micromobility, the real question is how the industry can win back trust and work with cities so the vision can be realized. Trust is only possible when cities and riders know that micromobility is not a fad, and that companies are no long swooping in with big promises and bigger valuations only to shutter overnight. We can't expect people to make fundamental changes to their transportation lifestyles (i.e., ditching on or all cars in cities) if companies are going to keep abandoning users when the money runs out. Similarly, infrastructure planning, design, and construction is not an overnight process. Cities cannot scrap their 20-year plans every time a new mobility service or company comes knocking. They have to know that they have a long-term, independently viable partner. That is where Veo comes in.

You don't exactly make headlines for being prudent and having foresight, and for that reason, many readers may not have even heard of Veo before now. But being trustworthy is what set us apart from the beginning, and what keeps us at the top of every city's shortlist today. We started small and listened intently as we grew. We listened to riders and cities and were able to innovate accordingly with our in-house design and manufacturing team. We kept our eye laser-focused on the long haul. We have zero cease and desist orders, only launching where we have been invited, and leaving graciously if cities need more time to build out safer infrastructure. Again, Veo stands alone in this position.

Veo is defining Micromobility 3.0, rooted in sustainability. I like to think of sustainability as a three-legged stool. To be around for the next century, we need to have financial sustainability, manufacturing sustainability, and rider sustainability by expanding the demographics of ridership.

To break down the first leg of the stool, financial practices need to clean up. Billions of dollars have been spent on off-the-shelf devices that flooded cities streets and ended up thrown off cliffs and into ditches. Without the infrastructure to support the new mode of transit, people felt assaulted with devices that were cheaper to trash than to repair. Veo never operated that way. In contrast, Veo was the first micromobility company to turn a profit. To do this, we focused on only launching in full partnership with cities. We quickly learned how to turn a profit on each individual program we built. While competitor's focused on fundraising and land grabbing, we focused on exceeding the expectations of the cities we were in and on learning the intricacies of operations, maintenance, and growth. Our vision was always to get it right on smaller scales first, rather than realizing the only thing you scaled was a mistake.

Speaking of quality, for the second leg of the stool, Veo designs and manufactures our vehicles in-house. This not only allows us to lead on key concerns like safety, but we are able to add new features and pressure test our gear before it is released to the rigors of shared use. Just last month we released a new version of the standing scooter that includes turn signals and an alarm to pick up the scooter from the right of way if it was knocked down. Both of these were the first in the industry and a result of listening to cities and riders. Veo also designed the swappable battery for shared scooters back in 2019—which have spread due to their utility and convenience.

Finally, and perhaps most importantly, micromobility companies must work with cities to widen the demographics of riders. We can do this by making devices feel safer for people of different physical abilities and preferences, providing more and different device options, and by making riding safe with protected lanes in parts of town that historically get looked over for innovations like this. While Veo already has a seated scooter option, the Cosmo, we are constantly innovating to fit the needs of more riders. We are also looking to solve for things that keep people from riding, like helmet head or inclement weather.

Cities know how to increase ridership on their end. Just as induced demand from highway expansions has shown that "If you build it, they will come"; this too, has been proven to apply to bike lanes, pedestrian centers, bike-share, and parks world-wide. In fact, many of us were delighted to see main streets shut down to cars to bring our favorite restaurants outside over the pandemic. My thought is keep those streets closed to cars. Let's edge out this unsustainable, unclean, and impersonal way of getting around by making it faster and cheaper to ride clean.

Progress seems to move slow when you're in the thick of it, but when you zoom out by a decade or three, it's amazing how fast the shifts occur. Let's give people the option that changed my life by making micromobility safe and sustainable. Finally, with Veo at the helm, Micromobility 3.0 is trustworthy enough to merit a foundational role. With Veo as a long-term partner, cities and riders can finally make long-term changes in how we travel and how cities build infrastructure. Together, we can focus on sustainability and let the transformation to utopia happen on its own.

About Candice Xie

Inspired by the challenges of creating a sustainable, alternative transportation system to meet first-mile/last-mile community needs with stellar products, Candice left corporate America to launch Veo in 2017. She knew success required building respectful relationships with communities and providing durable equipment designed exclusively for shared use. As CEO, Candice has grown Veo's business sustainably while balancing the safety and regulatory needs of communities with the unlock-and-go mentality of riders. Veo is the first profitable micromobility company, boasting financial sustainability unique to the industry. Under Candice's leadership, Veo is also the only company to design and manufacture bikes, ebikes, and seated and standing scooters in-house, giving the company a unique advantage to innovate on vehicle types and models specific to the needs of each market. Candice graduated with distinction in Finance from Purdue University.